Tangential Boundary Stabilization of Navier-Stokes Equations

MEMOIRS
of the
American Mathematical Society

Number 852

Tangential Boundary Stabilization of Navier-Stokes Equations

Viorel Barbu
Irena Lasiecka
Roberto Triggiani

May 2006 • Volume 181 • Number 852 (first of 5 numbers) • ISSN 0065-9266

American Mathematical Society
Providence, Rhode Island

2000 *Mathematics Subject Classification.*
Primary 76D05, 35B40, 35Q30.

Library of Congress Cataloging-in-Publication Data

Barbu, Viorel.
   Tangential boundary stabilization of Navier-Stokes equations / Viorel Barbu, Irena Lasiecka, Roberto Triggiani.
      p. cm. — (Memoirs of the American Mathematical Society, ISSN 0065-9266 ; no. 852)
   "Volume 181, number 852 (first of 5 numbers)."
   Includes bibliographical references.
   ISBN 0-8218-3874-1 (alk. paper)
   1. Navier-Stokes equations. 2. Boundary layer. 3. Mathematical optimization. 4. Riccati equation. I. Lasiecka, I. (Irena), 1948– II. Triggiani, R. (Roberto), 1942– III. Title. IV. Series.

QA3.A57 no. 852
[QA374]
510 s—dc22
[515′.353]   2006040678

Memoirs of the American Mathematical Society

This journal is devoted entirely to research in pure and applied mathematics.

Subscription information. The 2006 subscription begins with volume 179 and consists of six mailings, each containing one or more numbers. Subscription prices for 2006 are US$624 list, US$499 institutional member. A late charge of 10% of the subscription price will be imposed on orders received from nonmembers after January 1 of the subscription year. Subscribers outside the United States and India must pay a postage surcharge of US$31; subscribers in India must pay a postage surcharge of US$43. Expedited delivery to destinations in North America US$35; elsewhere US$130. Each number may be ordered separately; *please specify number* when ordering an individual number. For prices and titles of recently released numbers, see the New Publications sections of the *Notices of the American Mathematical Society*.

Back number information. For back issues see the *AMS Catalog of Publications*.

Subscriptions and orders should be addressed to the American Mathematical Society, P. O. Box 845904, Boston, MA 02284-5904, USA. *All orders must be accompanied by payment.* Other correspondence should be addressed to 201 Charles Street, Providence, RI 02904-2294, USA.

Copying and reprinting. Individual readers of this publication, and nonprofit libraries acting for them, are permitted to make fair use of the material, such as to copy a chapter for use in teaching or research. Permission is granted to quote brief passages from this publication in reviews, provided the customary acknowledgment of the source is given.

Republication, systematic copying, or multiple reproduction of any material in this publication is permitted only under license from the American Mathematical Society. Requests for such permission should be addressed to the Acquisitions Department, American Mathematical Society, 201 Charles Street, Providence, Rhode Island 02904-2294, USA. Requests can also be made by e-mail to reprint-permission@ams.org.

Memoirs of the American Mathematical Society is published bimonthly (each volume consisting usually of more than one number) by the American Mathematical Society at 201 Charles Street, Providence, RI 02904-2294, USA. Periodicals postage paid at Providence, RI. Postmaster: Send address changes to Memoirs, American Mathematical Society, 201 Charles Street, Providence, RI 02904-2294, USA.

© 2006 by the American Mathematical Society. All rights reserved.
This publication is indexed in *Science Citation Index*®, *SciSearch*®, *Research Alert*®, *CompuMath Citation Index*®, *Current Contents*®/*Physical, Chemical & Earth Sciences*.
Printed in the United States of America.

∞ The paper used in this book is acid-free and falls within the guidelines established to ensure permanence and durability.
Visit the AMS home page at http://www.ams.org/

10 9 8 7 6 5 4 3 2 1   11 10 09 08 07 06

Contents

Acknowledgements   ix

Chapter 1.   Introduction   1

Chapter 2.   Main results   13

Chapter 3.   Proof of Theorems 2.1 and 2.2 on the linearized system (2.4): $d = 3$ 21
   3.1.   Abstract models of the linearized problem (2.3). Regularity   21
   3.2.   The operator $D^* \mathcal{A}^*$, $D^* : H \to (L^2(\Gamma))^d$ 33
   3.3.   A critical boundary property related to the boundary c.c. in (3.1.2e)   35
   3.4.   Some technical preliminaries; space and system decomposition   36
   3.5.   Theorem 2.1, general case $d = 3$: An infinite-dimensional open-loop boundary controller g satisfying the FCC (3.1.22)–(3.1.24) for the linearized system (3.1.4): $g \in L^2(0, \infty; (L^2(\Gamma))^3)$, $g \cdot \nu \equiv 0$ on Σ; $y^0 \in (H^{\frac{1}{2}+\epsilon}(\Omega))^3 \cap H \Rightarrow y \in L^2(0, \infty; (H^{\frac{3}{2}+\epsilon}(\Omega))^3 \cap H)$ 38
   3.6.   Feedback stabilization of the unstable z_N-system (3.4.9) on Z_N^u under the FDSA   50
   3.7.   Theorem 2.2, case $d = 3$ under the FDSA: An open-loop boundary controller g satisfying the FCC (3.1.22)–(3.1.24) for the linearized system (3.1.4): $g \in L^2(0, \infty; (L^2(\Gamma_1))^d)$, $g \cdot \nu = 0$ on Σ, $g \equiv 0$ on $\Gamma \setminus \Gamma_1$; $y^0 \in (H^{\frac{1}{2}+\epsilon}(\Omega))^d \cap H \Rightarrow y \in L^2(0, \infty; (H^{\frac{3}{2}+2\epsilon}(\Omega))^d \cap H)$ 57

Chapter 4.   Boundary feedback uniform stabilization of the linearized system (3.1.4) via an optimal control problem and corresponding Riccati theory. Case $d = 3$ 61
   4.0.   Orientation   61
   4.1.   The optimal control problem (Case $d = 3$)   62
   4.2.   Optimal feedback dynamics: the feedback semigroup and its generator on W 64
   4.3.   Feedback synthesis via the Riccati operator   67
   4.4.   Identification of the Riccati operator R in (4.1.8) with the operator R_1 in (4.3.1)   76
   4.5.   A Riccati-type algebraic equation satisfied by the operator R on the domain $\mathcal{D}(A_R^2)$, where A_R is the feedback generator   80

Chapter 5.   Theorem 2.3(i): Well-posedness of the Navier-Stokes equations with Riccati-based boundary feedback control. Case $d = 3$ 83

Chapter 6. Theorem 2.3(ii): Local uniform stability of the Navier-Stokes equations with Riccati-based boundary feedback control   93

Chapter 7. A PDE-interpretation of the abstract results in Sections 5 and 6   95

Appendix A. Technical Material Complementing Section 3.1   97
  A.1. Extension of the Leray Projector P Outside the Space $(L^2(\Omega))^d$ 97
  Definition A.1.2: Extension of Leray projector to $H^{-1}(\Omega))^d$.   98
  A.2. Definition and Regularity of the Dirichlet Map in the General Case. Abstract Model   99

Appendix B. Boundary feedback stabilization with arbitrarily small support of the linearized system (3.1.4a) at the $(H^{\frac{3}{2}-\epsilon}(\Omega))^d \cap H$-level, with I.C. $y^0 \in (H^{\frac{1}{2}-\epsilon}(\Omega))^d \cap H$. Cases $d = 2, 3$. Theorem 2.5 for $d = 2$ 103
  B.1. An open-loop infinite dimensional boundary controller $g \in L^2(0, \infty; (L^2(\Gamma_1))^d)$, Γ_1 arbitrary, for the linearized system (3.1.4a) yielding $y(\,\cdot\,; y_0) \in L^2(0, \infty; (H^{\frac{3}{2}-\epsilon}(\Omega))^d \cap H)$ for $y^0 \in (H^{\frac{1}{2}-\epsilon}(\Omega))^d \cap H$ 103
  B.2. Feedback stabilization in $(H^{\frac{3}{2}-\epsilon}(\Omega))^d$, $d = 2, 3$, of the linearized system (3.1.4a) with finite-dimensional feedback controller with arbitrary support under FDSA = (3.6.2)   104
  B.3. Completion of the proof of Theorem 2.5 and Theorem 2.6 for the N-S model (1.1), $d = 2$ 106
  B.4. A regularity property of the Riccati operator corresponding to the linearized operator $\mathcal{A}$ in (1.11)   107

Appendix C. Equivalence between unstable and stable versions of the Optimal Control Problem of Section 4   113

Appendix D. Proof that $FS(\,\cdot\,) \in \mathcal{L}(W; L^2(0, \infty; (L^2(\Gamma))^d)$ 123

Bibliography   127

Abstract

The steady-state solutions to Navier-Stokes equations on a bounded domain $\Omega \subset R^d$, $d = 2, 3$, are locally exponentially stabilizable by a boundary closed-loop feedback controller, acting tangentially on the boundary $\partial\Omega$, in the Dirichlet boundary conditions. The greatest challenge arises from a combination between the control as acting on the boundary and the dimensionality $d = 3$. If $d = 3$, the non-linearity imposes and dictates the requirement that stabilization must occur in the space $(H^{\frac{3}{2}+\epsilon}(\Omega))^3$, $\epsilon > 0$, a high topological level. A first implication thereof is that, due to compatibility conditions that now come into play, for $d = 3$, the boundary feedback stabilizing controller *must* be infinite dimensional. Moreover, it generally acts on the entire boundary $\partial\Omega$. Instead, for $d = 2$, where the topological level for stabilization is $(H^{\frac{3}{2}-\epsilon}(\Omega))^2$, the boundary feedback stabilizing controller can be chosen to act on *an arbitrarily small* portion of the boundary. Moreover, still for $d = 2$, it may even be *finite* dimensional, and this occurs if the linearized operator is diagonalizable over its finite-dimensional unstable subspace.

In order to inject dissipation as to force local exponential stabilization of the steady-state solutions, an Optimal Control Problem (OCP) with a quadratic cost functional over an infinite time-horizon is introduced for the linearized N-S equations. As a result, the same Riccati-based, optimal boundary feedback controller which is obtained in the *linearized* OCP is then selected and implemented also on the full N-S system. For $d = 3$, the OCP falls definitely outside the boundaries of established optimal control theory for parabolic systems with boundary controls, in that the combined index of unboundedness—between the unboundedness of the boundary control operator and the unboundedness of the penalization or observation operator—is *strictly larger than* $\frac{3}{2}$, as expressed in terms of fractional powers of the free-dynamics operator. In contrast, established (and rich) optimal control theory [L-T.2] of boundary control parabolic problems and corresponding algebraic Riccati theory requires a combined index of unboundedness *strictly less than* 1. An additional preliminary serious difficulty to overcome lies at the outset of the program, in establishing that the present highly non-standard OCP—with the aforementioned high level of unboundedness in control and observation operators and subject, moreover, to the additional constraint that the controllers be *pointwise tangential*—be non-empty; that is, it satisfies the so-called Finite Cost Condition [L-T.2].

Received by the editor March 17, 2004.
2000 *Mathematics Subject Classification*. Primary: 76D05, 7655, 35B40, 35Q30.
Key words and phrases. Navier-Stokes equations, boundary feedback stabilization, optimal control, Riccati equation, steady-state solution, boundary feedback controller.

Acknowledgements

This work was performed, in part, during the visit of V. Barbu to the Mathematics Department of the University of Virginia, Fall 2002 and May–June 2003. The authors would like to thank the referees for much appreciated comments that led to an overall improvement of the presentation and exposition of the paper. In particular, in response to referees' queries, the authors have inserted a new appendix, the present Appendix A, to complement with more details the critical material of Section 3.1. A few pertinent additional references were also pointed out by the referees and duly included.

The research of I. L. and R. T. was partially supported by the National Science Foundation under grant NSF-DMS-0104305 and by the Army Research Office under grant ARO-DAAD19-02-1-0179. An announcement of the results of this paper were made by R.T. at the following conferences: (i) International Conference "Control of Partial Differential Equations," held at Georgetown University, May 29–June 2, 2003; (ii) 5th Congress of Romanian Mathematical Society, held at the University of Pitesti, Romania, June 22–28, 2003; (iii) the 21st IFIP Conference on System Analysis and Optimization, held at Sophia Antipolis, France, July 21–25, 2003; (iv) International Conference in Partial Differential Equations, held at Foz do Ignacu–PR Brazil, December 17–19, 2003. Short announcements of the main results of this paper have appeared in "Interior and Boundary Stabilization of the Navier-Stokes Equations," by R. Triggiani, *Bolletim de Sociedada Paranaence de Matematica* 22, Numero 1, 3 Serie (2004), 91–105; by R. Triggiani, in *System Modeling and Optimization*, edited by J. Cagnol & J. P. Zolesio, 2005, Kluwer Academic Publishers, pp. 41–57. Proceedings of the 21st IFIP-TC7 Conference, held July 21–25, 2003, in Sophia Antipolis, France; by V. Barbu, in *Control of Partial Differential Equations*, Chapman and Hall/CRC, Vol. 242, pp. 29–43 (2005), edited by O. Imannvilov,..., Editors. Proceedings of a conference held at Georgetown University, Washington, DC, May 30–June 1, 2003.

CHAPTER 1

Introduction

Boundary controlled Navier-Stokes equations. We consider the controlled Navier-Stokes equations (see [**C-F.1**, p. 45], [**Te.1**, p. 253] for the uncontrolled case $u \equiv 0$) with boundary control u in the Dirichlet B.C.:

$$
\begin{aligned}
(1.1\text{a}) && y_t(x,t) - \nu_0 \Delta y(x,t) + (y \cdot \nabla) y(x,t) &= f_e(x) + \nabla p(x,t) && \text{in } G; \\
(1.1\text{b}) && \nabla \cdot y &= 0 && \text{in } G; \\
(1.1\text{c}) && y &= u && \text{on } \Sigma; \\
(1.1\text{d}) && y(x,0) &= y_0(x) && \text{in } \Omega.
\end{aligned}
$$

Here, $G = \Omega \times (0, \infty)$; $\Sigma = \partial\Omega \times (0, \infty)$ and Ω is an open smooth bounded domain of R^d, $d = 2, 3$; $u \in L^2(0, T; (L^2(\partial\Omega))^d)$ is the boundary control input; and $y = (y_1, y_2, \ldots, y_d)$ is the state (velocity) of the system. The constant $\nu_0 > 0$ is the viscosity coefficient. The functions $y_0, f_e \in (L^2(\Omega))^d$ are given, the latter being a body force, while p is the unknown pressure. Because of the divergence theorem: $\int_\Omega \nabla \cdot y \, d\Omega = \int_\Gamma y \cdot \nu \, d\Gamma$, $[\Gamma = \partial\Omega$, $\nu =$ unit outward normal to $\partial\Omega]$, we must require (at least) the integral boundary compatibility condition: $\int_\Gamma u \cdot \nu \, d\Gamma = 0$ on the control function u. Actually, a more stringent condition has to be imposed, in our final results: $u \cdot \nu \equiv 0$ on Σ, to sustain the pointwise boundary compatibility condition contained in the definition of the critical state space H in (1.5a) below. To summarize, we shall then assume

$$
(1.1\text{e}) \qquad \text{either } u \cdot \nu \equiv 0 \text{ on } \Sigma; \text{ or at least } \int_\Gamma u \cdot \nu \, d\Gamma \equiv 0, \text{ a.e. } t > 0,
$$

as it will be specified on a case-by-case basis.

Steady-state solutions. Let $(y_e, p_e) \in ((H^2(\Omega))^d \cap V) \times H^1(\Omega)$ be a steady-state solution to equation (1.1), i.e.,

$$
(1.2) \qquad \begin{cases} -\nu_0 \Delta y_e + (y_e \cdot \nabla) y_e &= f_e + \nabla p_e && \text{in } \Omega; \\ \nabla \cdot y_e &= 0 && \text{in } \Omega; \\ y_e &= 0 && \text{on } \partial\Omega. \end{cases}
$$

The steady-state solution is known to exist for $d = 2, 3$, [**C-F.1**, Theorem 7.3, p. 59]. Here [**C-F.1**, p. 9], [**Te.1**, p. 18]

$$
(1.3) \qquad V = \{ y \in (H_0^1(\Omega))^d; \, \nabla \cdot y = 0 \},
$$

$$
\text{with norm } \|y\|_V \equiv \|y\| = \left\{ \int_\Omega |\nabla y(x)|^2 d\Omega \right\}^{\frac{1}{2}}.
$$

Goal. For large Reynolds numbers $\frac{1}{\nu_0}$, the steady-state (stationary) solutions y_e are unstable and cause turbulence in their surroundings. Let y_e be one such unstable steady-state solution. Our goal is then to construct a boundary control u, subject to the boundary compatibility condition (c.c.) given by (1.1e) in the strong pointwise form $u \cdot \nu \equiv 0$ on $\partial\Omega$, and, moreover, in feedback form $u = \mu(y - y_e)$ (where $y_e|_{\partial\Omega} = 0$ by (1.2)) via some linear operator μ: $y - y_e \to u$, such that, once $u = \mu(y - y_e)$ is substituted in (1.1c), the resulting well-posed, closed-loop system (1.1a–d) possesses the following desirable property: *the steady-state solution y_e defined in (1.2) is locally exponentially stable*. This means that such feedback solution y initiating near the unstable y_e approaches asymptotically y_e. Once this is achieved, one has therefore managed to suppress turbulence asymptotically, by means of a closed-loop boundary control. In particular, motivated by our prior effort [**B-T.1**] to be described below, we seek to investigate if and when the feedback controller $u = \mu(y - y_e)$ can be chosen to be finite-dimensional, and, moreover, to act on an arbitrarily small portion (of positive measure) of the boundary $\Gamma = \partial\Omega$. A controller u is here called *finite-dimensional* in case it is the form $u = \sum_{i=1}^{k} u_i(t) w_i(x)$. Otherwise, the controller u will be called infinite-dimensional.

Orientation. Use of the Optimal Control Problem and Algebraic Riccati Theory. ($d = 3$) We emphasize here only the more demanding case of $d = 3$. A preliminary difficulty (for $d = 2, 3$) is the requirement that we impose on (1.1e) that the boundary control u be always tangential at each point of the boundary. It is standard that this requirement is intrinsically built in the definition of the state space H (below in (1.5a)) of the velocity vector y, which is critical to eliminate the second unknown of the N-S model, the pressure term ∇p (see the orthogonal complement $H^\perp$ in (1.5b) below), by virtue of the Leray projection P below. Evolution of the velocity must occur in H. Accordingly, we must then have that the boundary controls be pointwise tangential: $u \cdot \nu \equiv 0$ on Σ in (1.1e). Next, a second difficulty, this time for $d = 3$, is that the non-linearity of the N-S equation dictates and forces the requirement that stabilization must occur in the space $(H^{\frac{3}{2}+\epsilon}(\Omega))^3$, $\epsilon > 0$, see Eqn. (5.18a–b) below. This is a high topological level, of which we shall have to say more below. A third source of difficulty consists in deciding how to inject 'dissipation' into the N-S model, in fact, as required, through *a boundary* tangential controller expressed in feedback form. Here, motivated by [**B-T.1**] and, in turn, by optimal control theory [**L-T.1**], [**L-T.2**], in order to inject dissipation into the N-S system as to force local exponential boundary stabilization of its steady-state solutions, we choose the strategy of introducing an Optimal Control Problem (OCP) with a quadratic cost functional, over an infinite time-horizon, for the *linearized* N-S model subject to tangential Dirichlet-boundary control u, i.e., satisfying $u \cdot \nu \equiv 0$ on Σ. One then seeks to express the boundary feedback, closed-loop controller of the optimal solution of the OCP, in terms of the Riccati operator arising in the corresponding algebraic Riccati theory. As a result, the same Riccati-based boundary feedback optimal controller that is obtained in the *linearized* OCP is then selected and implemented also on the full N-S system. This controller in feedback form is both dissipative as well as 'robust' (with respect to a certain class of perturbations). For $d = 3$, however, the OCP must be resolved at the *high* $(H^{\frac{3}{2}+\epsilon}(\Omega))^3$-topological level, within the class of Dirichlet *boundary* controls in $L^2(0, \infty; (\partial\Omega)^3)$, which are further constrained to be *tangential to the boundary*.

Thus, the OCP faces two additional difficulties that set it apart and definitely outside the boundaries of established optimal control theory for parabolic systems

1. INTRODUCTION

with boundary controls: (1) the high degree of unboundedness of the *boundary* control operator, of order ($\frac{3}{4}+\epsilon$) as expressed in terms of fractional powers of the basic free-dynamics generator; and (2) the high degree of unboundedness of the 'penalization' or 'observation' operator of order also ($\frac{3}{4}+\epsilon$), as expressed in terms of fractional powers of the basic free-dynamics generator. This yields a 'combined index' of unboundedness *strictly greater than* $\frac{3}{2}$. By contrast, the established (and rich) optimal control theory of boundary control parabolic problems and corresponding algebraic Riccati theory requires a 'combined index' of unboundedness *strictly less than* 1 [**L-T.2**, Vol. 1, in particular, p. 501–503], which is the maximum limit handled by perturbation theory of analytic semigroups. To implement this program, however, one must first overcome, at the very outset, the preliminary stumbling obstacle of showing that the present highly non-standard OCP—with the aforementioned high level of combined unboundedness in control and observation operators and further restricted within the class of *tangential* boundary controllers—is, in fact, *non-empty*. This result is achieved in Theorem 3.5.1 in full generality (and in Proposition 3.7.1, in a more explicit and desirable form, under the assumption that the linearized operator is diagonalizable over its finite-dimensional unstable subspace). Thus, after this result, the study of the OCP may then begin. Because of the aforementioned intrinsic difficulties of the OCP with a combined index of unboundedness $> \frac{3}{2}$, one cannot (and cannot hope to) recover in full all desirable features of the corresponding algebraic Riccati theory which are available when the combined index of unboundedness in control and observation operators is *strictly less than* 1 ([**L-T.2**] and references therein). For instance, existence of a solution (Riccati operator) of the algebraic Riccati equation is here asserted only on the domain $\mathcal{D}(A_R^2)$ of the operator A_R^2, where A_R is the generator of the optimal feedback dynamics (Proposition 4.5.1 below); not on the domain of the free-dynamics operator, as it would be required by, or at least desirable from, the viewpoint of the OCP. However, in our present tratment, the OCP is a means to extract dissipation and stability, not an end in itself. And indeed, the present study of the algebraic Riccati theory, with a combined index of unboundedness in control and observation operator *strictly above* $\frac{3}{2}$ (rather than *strictly less than* 1) does manage, at the end, to draw out the key sought-after features of interest—dissipativity and decay—for the resulting optimal solution in feedback form of the OCP for the linearized N-S equation. All this is accomplished in Section 4.

The subsequent step of the strategy is then to select and use the same Riccati-based, boundary feedback operator, which was found to describe the optimal solution of OCP of the *linearized* N-S equation, directly into the full N-S model. For $d = 3$, the heavy groundwork for the feedback stabilization of the linearized problem via optimal control theory makes then the resulting analysis of well-posedness (in Section 5) and stabilization (in Section 6) of the N-S model more amenable than would otherwise be the case.

To this end, key use is made of the Algebraic Riccati Equation (ARE) satisfied by the Riccati operator that describes the stabilizing control in closed-loop feedback form, by means of the feedback operator F of Proposition 4.3.4. But, throughout Section 4, such operator F can only be defined on $\mathcal{D}(A_R^2)$; and it is only on $\mathcal{D}(A_R^2)$

that the ARE (4.5.1) of the OCP is asserted to yield the Riccati operator R_1, that—after a critical identification, via the optimization problem, to boost its regularity—is then the key operator that defines the feedback operator F. Thus, two more crucial problems remain to be addressed:

(1) First, extend the operator F, at least on the feedback dynamics $S(t)x$, generated by A_R, for *any* x in the state space W, not just in $\mathcal{D}(A_R^2)$, so that $FS(t)x \in L^2(0,\infty;(L^2(\Gamma))^3)$: this is achieved in Appendix D, via the preliminary estimate (5.7) of Lemma 5.3 valid initially (part (i), Eqn. (5.10)) for $x \in \mathcal{D}(A_R^2)$, and later extended to all $x \in W$ by Appendix D.

(2) Next, the PDE-interpretation of the abstract results of Sections 4, 5, 6, and Appendix D; this is discussed in the final Section 7, by making key use of the preliminary interpretation of F as $F = \nu_0 \frac{\partial}{\partial \nu} R$ on $\mathcal{D}(A_R^2)$, along with the extension result of Appendix D.

Literature. This paper is a successor to [**B-T.1**], which instead considered the *interior* stabilization problem of the Navier-Stokes equations, that is, problem (1.1) with (i) non-slip boundary condition $y \equiv 0$ on $\Sigma \equiv \partial\Omega \times (0,\infty)$ in place of the boundary controlled condition (1.1c); and (ii) interior control $m(x)u(x,t)$ on the right-hand side of Eqn. (1.1a), where $m(x)$ is the characteristic function of an *arbitrary* open subset $\omega \subset \Omega$ of positive measure. In this case, [**B-T.1**] proves that (the linearized problem is exponentially stabilizable, hence that) the steady-state solutions y_e to the Navier-Stokes equations are locally exponentially stabilizable by a *finite-dimensional* feedback controller, in fact, *of minimal size*. In addition, one may select the *finite-dimensional* feedback controller to be expressed in terms of a Riccati operator (solution of an algebraic Riccati equation, which arises in an optimization problem associated with the linearized equation). *We shall need to invoke this interior stabilization problem* (though not in its full strength) *in Section 3.5, in order to solve the present boundary stabilization problem.*

The work in the literature which is most relevant to our present paper is that of A. Fursikov, see [**F.1**], [**F.2**], [**F.3**] (of which we became aware after completing [**B-T.1**]), which culminates a series of papers quoted therein. A statement of the main contribution of [**F.1**], as it pertains to the linearized problem (2.3) below, is contained in [**F.1**, Theorems 3.3 and 3.5, pp. 104–5]. Given the pair $\{\Omega,\Gamma_0\}$, where Γ_0 is a portion of the boundary $\partial\Omega$ of Ω, the approach of [**F.1**] starts with a special class, called $V^1(\Omega,\Gamma_0)$, of initial conditions y^0 for the linearized problem (2.3) defined on Ω. More specifically, $y^0 \in V^1(\Omega,\Gamma_0)$ means that the initial vector y^0 is the restriction $y^0 = Y^0|_\Omega$ on Ω of the vector Y^0 on the extended domain $G = \Omega \cup \omega$ (in the notation of [**F.1**]), where: (i) $y^0 \in (H^1(G))^d$, $d = 2,3$; (ii) $\nabla \cdot Y^0 \equiv 0$ in G; (iii) $Y^0|_{\partial G} = 0$; (iv) the set ω is an extension of Ω across the pre-assigned portion Γ_0 of $\partial\Omega$. Then, [**F.1**, Theorem 3.3] establishes that such $y^0 \in V^1(\Omega,\Gamma_0)$ can, in turn, be re-extended from Ω to G as a new vector $\eta^0 \in (H^1(G))^d$, $\nabla \cdot \eta^0 \equiv 0$ in G, $\eta^0|_{\partial G} = 0$, possessing now the new critical property that, in addition, η^0 belongs to the *stable* subspace of the linearized operator $\tilde{\mathcal{A}}$. Here $\tilde{\mathcal{A}}$ is an extension of the original operator $\mathcal{A}$ in (1.11) from Ω to G, obtained by a corresponding extension of the steady-state solution y_e from Ω to G, while preserving the required properties of being divergence-free across G and vanishing on ∂G. As a consequence of this extension, one obtains a function $\eta(t;\eta^0) = e^{\tilde{\mathcal{A}}t}\eta^0$, which is the solution of the corresponding linearized problem, except this time on G, and with homogeneous Dirichlet (non-slip) boundary condition on ∂G. Then, [**F.1**, Theorem 3.5] concludes (because of the obvious invariance of the stable subspace for the s.c. analytic semigroup $e^{\tilde{\mathcal{A}}t}$) that such $\eta(t;\eta^0)$ is exponentially decaying: $\|e^{\tilde{\mathcal{A}}t}\eta^0\| \leq Ce^{-\sigma t}\|\eta^0\|$, in the $(H^1(G))^d$-norm $\|\ \|$, with a controlled decay rate σ, $0 < \sigma < |\text{Re } \lambda_{N+1}|$, λ_{N+1} being the first unstable eigenvalue of $\mathcal{A}$. At this point, [**F.1**] takes the restrictions of $\eta(t;\eta^0)$ on Ω and Γ_0: $y(t) = \eta(t;\eta^0)|_\Omega$, $u(t) = \eta(t;\eta^0)|_{\Gamma_0}$, with $u(t) \equiv 0$ on $\Gamma \setminus \Gamma_0$, and defines such u as a "feedback control" of the corresponding solution y of problem (2.3) on Ω, which then stabilizes such solution $y(t)$ of (2.3) over Ω. A similar definition of "feedback control" is given in [**F.1**] with respect to the non-linear Navier-Stokes problem.

One should note, however, that the aforementioned controller for problem (2.3) given in [**F.1**] is not a feedback controller in the standard sense. Instead, our main results (in [**B-T.1**] as well as) in the present paper construct genuine, authentic, and real feedback controls (Riccati-based, in fact, hence with some feature of 'robustness'), that use at time t only the state information on Ω at time t. The present paper, therefore, encounters a host of technical problems not present in [**F.1**]: from the need for the genuine feedback control u to satisfy the pointwise compatibility condition $u \cdot \nu \equiv 0$ on Σ; to the high topological level $(H^{\frac{3}{2}+\epsilon}(\Omega))^3$ at which stabilization must occur in our case, as dictated by the non-linearity for $d = 3$, see Eqn. (5.18a–b), versus the H^1-topology decay obtained in [**F.1**]; to the treatment of the Riccati theory for a corresponding optimal control problem with a combined 'index of unboundedness' in control and observation operators exceeding $\frac{3}{2}$ —thus $\frac{1}{2} + 2\epsilon$ beyond the (rich) theory of the literature [**L-T.2**], as explained in the preceding *Orientation*.

In closing, we recall that—alternatively—given an unstable steady-state solution y_e and a preassigned time $T > 0$, one may seek an *open-loop* boundary control u such that the corresponding solution y initiating near y_e hits y_e at the time T: $y(T) = y_e$. This problem is referred to as *local exact boundary controllability*. It is solved in [**F-I.1**], [**F-I.2**] for $d = 2, 3$, by virtue of Carleman estimates for elliptic and inverse parabolic equations.

Main contributions of the present paper. Qualitative summary of main results. A first qualitative description of the main results of the present paper follows next. First of all, the pre-set goal is achieved: *with no assumptions whatsoever (except mild assumptions on the domain), we prove here that the steady-state solutions to Navier-Stokes equations on $\Omega \subset \mathbb{R}^d$, $d = 2, 3$, are locally exponentially stabilizable by a closed-loop boundary feedback controller acting in the Dirichlet boundary conditions in the required topologies [Theorem 2.3 for $d = 3$ and Theorem 2.5 for $d = 2$]*. The feedback controller is expressed in terms of a Riccati operator (solution of a suitable algebraic Riccati equation): as such, via standard arguments (e.g., [**B-T.1**, Remark 4.2, p. 1483]) this feedback controller is 'robust' with respect to a certain class of exogeneous perturbations.

More precisely, the following main results are established in the present paper:

(i) For the general cases $d = 2, 3$, an infinite-dimensional, closed-loop boundary feedback, stabilizing controller is constructed, as acting (for general initial data) on the entire boundary $\partial\Omega$ for $d = 3$; or on an arbitrarily small portion of the boundary, for $d = 2$.

(ii) By contrast, for $d = 2$ and under a finite-dimensional spectral assumption FDSA = (3.6.2) (diagonalizability of the restriction of the linearized operator over the finite-dimensional unstable subspace), the feedback controller can be chosen to be *finite-dimensional*, with dimension related to properties of the unstable eigenvalues, and, moreover, still to act on an *arbitrarily small portion* of the boundary.

(iii) For $d = 3$, local exponential feedback stabilization of the steady-state solutions to Navier-Stokes equations is *not* possible with a *finite-dimensional* boundary feedback controller (except for a meager set of special initial conditions).

(iv) The pathology noted in (iii) for $d = 3$ is due to the non-linearity (see Eqn. (5.18a–b)) which (via Sobolev embedding and multiplier theory [**M-S.1**], [**R-S.1**] for $d = 3$) forces then the requirement that solutions of the linearized problem be considered at the high regularity space $(H^{\frac{3}{2}+\epsilon}(\Omega))^d \cap H$, $\epsilon > 0$, under initial conditions $y_0 \in (H^{\frac{1}{2}+\epsilon}(\Omega))^d \cap H$ and $L^2(0, \infty; (L^2(\Gamma))^d)$-boundary controls u. In turn, this high regularity space $H^{\frac{3}{2}+\epsilon}(\Omega)$ causes the occurrence of the compatibility condition $y_0|_\Gamma = u(0)$ at $t = 0$ on the boundary to be satisfied. Thus, for $d = 3$, the constructed feedback controller must be infinite-dimensional in general (see Proposition 3.1.3).

(v) By contrast, the *linearized* problem for $d = 2, 3$ is exponentially stabilizable with a closed-loop boundary, *finite-dimensional* feedback-controller acting on an *arbitrarily small portion* of the boundary up to the topological level $(H^{\frac{3}{2}-\epsilon}(\Omega))^d$ and with initial conditions $y_0 \in (H^{\frac{1}{2}-\epsilon}(\Omega))^d \cap H$, under the same FDSA = (3.6.2).

Technical preview of issues resolved. The above Orientation aimed at extracting the technical setting of the problem along with a description of the main conceptual difficulties encountered in its solution. Here, we follow up with a more technical outline of how the aforementioned obstacles have been overcome. In order to achieve the final sought-after goal—the boundary stabilization of the 3-d Navier-Stokes equations, as stated in Theorem 2.3—the present treatment had to overcome a series of novel issues, each of which possessing an independent interest of its own and having an impact on other related problems. Accordingly, we feel that it is worthy for us to extract and single out these building blocks in our treatment. The contributions of the present article include solutions to the following sub-problems of wider interest, beyond the present paper:

(1) Formulation of an abstract, semigroup based model to describe first the linearized N-S mixed problem (2.3) (or (3.1.2)) and next the full N-S problem (1.1) (or (2.1)), with forcing term (control function) acting on the boundary $\partial\Omega$ of the spatial domain (in any dimension) in the Dirichlet boundary conditions. This model—Eqn. (3.1.4a) for the linearized N-S problem and (5.0) for the fully N-S problem—while in line with past results on abstract modeling of boundary control problems for PDEs (parabolic, hyperbolic, Petrowski-type, coupled systems of PDEs, etc.), see [**L-T.1**], [**L-T.2**], has to critically take into consideration novel features, or requirements, of N-S problems. These must include the well-known constraint that the natural state space for the velocity vector is the space H in (1.5a) below, equipped with the divergence-free condition and the pointwise boundary compatibility condition $g \cdot \nu \equiv 0$ in (1.1e), in order to eliminate, by Leray projection, the second unknown of the N-S model, the pressure term ∇p. Our abstract model of the velocity is accordingly given on the space H, in which case there exists a critical trace interpretation given by Proposition 3.2.1. This point is handled in Sections 3.1, 3.2, 3.3, complemented by Appendix A.

(2) Optimal regularity theory particularly of the mixed (initial/boundary value) problem of the linearized N-S abstract model over a broad topological range of data, with or without compatibility relations. This phase uses to advantage the model in (1) plus PDE techniques. This treatment begins with Theorem 3.1.4, with boundary datum $g \in H^{0,0}(\Sigma)$. This is obtained by interpolation between the higher level case of $g \in H^{\frac{3}{2},\frac{3}{4}}(\Sigma)$, subject to a compatibility relation with the initial condition on the one hand (Lemma 3.1.5), and the lower-level case of $g \in [H^{\frac{1}{2},\frac{1}{4}}(\Sigma)]'$ (Lemma 3.1.6), on the other hand, which is carried out by duality. This optimal regularity theory then culminates with the general result of Theorem 3.1.8 with boundary datum $g \in H^{2\theta,\theta}(\Sigma)$, $0 \leq \theta \leq 1$, subject to a compatibility relation with the initial condition, wherever applicable. The case of key interest upon which the main treatment of the paper rests for $d = 3$ is obtained for $2\theta = 1 + \epsilon$ (Proof of Theorem 3.5.1; in particular of the regularity in (3.5.59)): this critical value falls on a range of the interpolation parameter, which we have *not* found included even in thorough treatments of *classical* parabolic equations such as [**L-M.1**, Vol. 2, see p. 79, middle and p. 87]. In particular, refer to Remark 3.3.1 below. For other ranges of the interpolation parameter, the theory presented here for the linearized N-S problem is consistent with the optimal theory of *classical* parabolic operators [**L-M.1**, Vol. 2].

(3) Proof, for $d = 3$, that the quadratic cost, optimal control problem (OCP) over an infinite horizon of the *linearized* N-S problem (2.3) [or (3.1.2) or, in abstract form, (3.1.4a), as in point (1)] is *not* empty (or "proper" in the language of optimization theory). This means establishing the following property for the linearized N-S problem (3.1.4a): Given

(I.C.) any initial condition $y^0 \in (H^{\frac{1}{2}+\epsilon}(\Omega))^3 \cap H$,

there always exists a Dirichlet boundary datum

(B.C.) $g \in L^2(0,\infty;(L^2(\partial\Omega))^3)$, subject to further pointwise c.c. $g \cdot \nu \equiv 0$ on $\Gamma \times (0,\infty)$,

such that the corresponding solution y of the linearized N-S equation (3.1.4a) satisfies the interior regularity property

(I.R.) $y \in L^2(0,\infty;(H^{\frac{3}{2}+\epsilon}(\Omega))^3 \cap H) \cap H^{\frac{3}{4}+\frac{\epsilon}{2}}(0,\infty;H)$.

We note the peculiar and 'non-standard' topological setting of the above statement. As unusual and unattractive as it may look, this topological setting of the *linearized* N-S equation (3.1.4a) is, however, the one that is *dictated and imposed by the non-linearity* of the full N-S problem (1.1) or (2.1), for $d = 3$ (see Eqn. (5.18b)) as already stressed before. The above regularity result (I.C.) + (B.C.) $\Rightarrow$ (I.R.) is non-trivial. It is established in Theorem 3.5.1. One ingredient of its proof is the optimal regularity theory of the mixed linearized N-S problem of point (2); more specifically, the aforementioned Theorem 3.1.8 for the value θ of the interpolating parameter satisfying $2\theta = 1 + \epsilon$, see (3.5.59) (in a range, as we have already noted, where a corresponding optimal regularity theory for even *classical* parabolic operators is *not* to be found in thorough established treatments such as [**L-M.1**, Vol. 2]). Another source of difficulty is the peculiar class of initial conditions in (I.C.), non-natural even to determine the action on it, or behavior from it, of the corresponding linearized N-S semigroup. A further serious obstacle is represented by the necessity of establishing that the sought-after Dirichlet boundary datum g satisfies also the required pointwise boundary compatibility condition $g \cdot \nu \equiv 0$ on $\Gamma \times (0, \infty)$. These and other difficulties are taken up in Theorem 3.5.1, which establishes the validity of the implication (I.C.) + (B.C.) $\Rightarrow$ (I.R.).

To handle these difficulties, we extend the original initial condition y^0 in the (unnatural) topology $y^0 \in (H^{\frac{1}{2}+\epsilon}(\Omega))^3 \cap H$ over the original domain Ω to an extended $\tilde{y}_0$ in a (natural) topology $\tilde{y}_0 \in (H_0^{\frac{1}{2}+\epsilon}(\tilde{\Omega}))^3 \cap \tilde{H}$ on the extended domain $\tilde{\Omega} = \Omega \cup \omega$, which is natural for the action on it of the extended linearized N-S semigroup on $\tilde{\Omega}$. Next, a recent result [**B-T.1**] of uniform stabilization of the linearized N-S equations with interior controls with support localized on the strip ω (outside to Ω) of $\tilde{\Omega}$ is critically invoked to obtain a solution $\tilde{y}$ of the N-S linearized problem over $\tilde{\Omega}$, which is uniformly (exponentially) decaying on $\tilde{\Omega}$. We then take the restriction $\tilde{y}|_\Gamma$ of $\tilde{y}$ on the boundary $\Gamma = \partial\Omega$. But such restriction has the deficiency of yielding $\tilde{y}|_\Gamma \cdot \nu \not\equiv 0$ and hence of non-satisfying the required pointwise boundary c.c. in (3.1.2e) stressed in (1). Thus, it needs to be modified by subtracting its normal components, thus obtaining $\tilde{y}|_\Gamma - (\tilde{y}|_\Gamma \cdot \nu)\nu$ as a required tangential boundary term. But this changes then the solution over $\tilde{y}$ in Ω. Is the new solution still stable in Ω? This question raises additional difficulties to be overcome. At the end, Lemma 3.3.1 is employed to guarantee that the new boundary term $g = (\tilde{y} \cdot \nu) - (\tilde{y} \cdot \nu) \cdot \nu$, $g \cdot \nu \equiv 0$ is the sought-after boundary datum, producing a stable solution in Ω.

(4) Solution of the Optimal Control Problem of the linearized N-S Dirichlet *boundary* model (3.1.4a)—referred to in point (3) whose role was amply described in the Orientation—at the required *high* topological level of $(H^{\frac{3}{2}+\epsilon}(\Omega))^3 \cap H$, including, of course, the corresponding Algebraic Riccati Equations (ARE) Theory for the closed-loop feedback form of the optimal solution. This treatment, carried out in Section 4, represents a definite advance over the rather comprehensive, rich theory of OCP and Algebraic Riccati Theory for abstract parabolic problems, already available in the literature [**L-T.1**], [**L-T.2**]. This is so since such literature covers boundary control for parabolic or parabolic-like PDEs (or their corresponding abstract models) up to the maximum perturbation theory limit, which is obtained when the combined index of unboundedness of control and observation operators is *strictly less than* 1, as measured in terms of fractional powers of the free-dynamics generator. By contrast, as already noted in the Orientation, the present N-S problem features a combined index of unboundedness strictly greater than $\frac{3}{2}$ (that is, $\frac{3}{4} + \epsilon$ for the control operator and $\frac{3}{4} + \epsilon$ for the observation operator). This extension, though inspired by previous literature [**L-T.1**], [**L-T.2**], [**F-L-T.1**], represents a new technical achievement. It is not aimed at a generalization across the full range of the corresponding Algebraic Riccati Equation theory—which is simply not true—but rather on selected ingredients of interest here: the uniform stabilization of the optimal solution with a Riccati-based, closed loop feedback form. It is conducted largely at the abstract level, with an infusion, however, of special features tied to the underlying dynamics of the linearized N-S equations. Accordingly, it may prove useful for additional non-standard parabolic situations, requiring an OCP/Algebraic Riccati Equations Theory above the aforementioned limit of 1 for the combined index of unboundedness of control and observation operators. A noteworthy ingredient to be henceforth invoked for any treatment of OCP/ARE over an infinite horizon is given in Appendix C, which provides two different proofs for the following quite useful result: that the OCP/ARE theory in the case of an *unstable*

free dynamics operator can always be reduced to a corresponding treatment with a *stable* free dynamics operator.

(5) Analysis of the closed-loop full N-S model that results by expressing the boundary stabilizing control in terms of the *same* Riccati-based feedback form obtained in the OCP of the *linearized* N-S equations. This includes: (a) first, global well-posedness for sufficiently small I.C. y^0 in the right W-norm, $W \equiv (H^{\frac{1}{2}+\epsilon}(\Omega))^3 \cap H$, of the resulting closed-loop full N-S model by virtue of energy methods and fixed points. This is critically based, however, on the Algebraic Riccati Equation of the OCP which is satisfied by the Riccati operator describing the stabilizing feedback controller (Theorems 5.1 and 5.2, and Appendix D); (b) finally, local uniform stabilization of the full N-S model with Dirichlet boundary control, as in Theorem 2.3 (Theorem 6.1 of Section 6), by virtue of energy methods and semigroup methods.

Notation and preliminaries. Here we shall use the standard notation for the spaces of summable functions and Sobolev spaces on Ω and Γ. In particular, $H^s(\Omega)$ is the Sobolev space of order s with the norm denoted by $\|\cdot\|_s$. The following notation will be also used:

(1.4) $\quad \nabla \cdot y = \text{div } y, \ (y \cdot \nabla)y = y_i D_i y_j = y \cdot \nabla y_j, \ j = 1, \ldots, d, \ D_i = \dfrac{\partial}{\partial x_i};$

(1.5a) $\quad H = \{y \in (L^2(\Omega))^d; \ \nabla \cdot y = 0, \ y \cdot \nu = 0 \text{ on } \partial\Omega\} \quad$ [**C-F.1**, p. 7];

(1.5b) $\quad H^\perp = \{y \in (L^2(\Omega))^d : y = \text{grad } p, \ p \in H^1(\Omega)\}, \ (L^2(\Omega))^d = H + H^\perp,$

$H^\perp$ being the orthogonal complement of H in $(L^2(\Omega))^d$ [**Te.1**, p. 15], with summation convention to be used throughout the paper, presently in $i = 1, \ldots, d$, where ν is the unit outward normal to the boundary $\partial\Omega$ of Ω. We shall denote by $(\cdot, \cdot)$ the scalar product in both H and $(L^2(\Omega))^d$. Similarly, we shall denote by the same symbol $|\cdot|$ the norm of both $(L^2(\Omega))^d$ and H, and by $\|\cdot\|$ the norm of the space V as defined in (1.3).

We shall denote by $P : (L^2(\Omega))^d \to H$ the Leray projector [**C-F.1**, p. 9], which is orthogonal on $(L^2(\Omega))^d$. Thus, applying P to Eqn. (1.1a) eliminates the pressure term on H by virtue of (1.5). Furthermore, we have $Py_t = y_t$, provided that $y \in H$, i.e., provided that the pointwise c.c. $u \cdot \nu = 0$ on Σ is imposed; see also Section A.1 of Appendix A. In addition, we set

(1.6) $\quad Ay = -P\Delta y, \ \forall y \in \mathcal{D}(A) = (H^2(\Omega))^d \cap V, \quad V = \mathcal{D}(A^{\frac{1}{2}}),$

which is a self-adjoint positive definite operator in H with compact (resolvent) A^{-1} on H [**C-F.1**, p. 32]. Accordingly, the fractional powers A^s, $0 < s < 1$, are well-defined [**C-F.1**, p. 33]. We have $V = \mathcal{D}(A^{\frac{1}{2}})$ [**C-F.1**, p. 33]. Furthermore, we define $B : V \to V'$ by [**C-F.1**, p. 54], [**Te.1**, p. 162],

(1.7) $\quad By = P[(y \cdot \nabla)y], \quad (By, w) = b(y, y, w), \quad \forall y, w \in V,$

where the trilinear form is defined by [**C-F.1**, p. 49], [**Te.1**, p. 161]

(1.8) $\quad b(y, z, w) = \displaystyle\int_\Omega y_i(D_i z_j) w_j dx = \int_\Omega \langle y \cdot \nabla z, w \rangle_{\mathbb{R}^d} d\Omega, \quad y, w \in H, \ z \in V.$

The trilinear form b in (1.8), describing the non-linearity, obeys the following well-known estimate [**C-F.1**, p. 50], to be repeatedly invoked in the sequel:

(1.9) $\quad |b(y, z, w)| \leq C \|y\|_{m_1} \|z\|_{m_2+1} \|w\|_{m_3},$

where $\|\cdot\|_m$ is, as said, the norm of the Sobolev space $(H^m(\Omega))^d$ and the non-negative subindexes m_i are assumed to satisfy the constraints [**C-F.1**, p. 49]:
  (i) either $m_1 + m_2 + m_3 \geq \frac{d}{2}$, if $m_i \neq \frac{d}{2}$ for all $i = 1, 2, 3$;
  (ii) or else $m_1 + m_2 + m_3 > \frac{d}{2}$, if $m_i = \frac{d}{2}$ for at least one i.

Finally, in order to describe the linearized problem (2.3) below, we shall introduce the operator $A_0 \in \mathcal{L}(V; H)$,

(1.10a) $\qquad A_0 y = P((y_e \cdot \nabla)y + (y \cdot \nabla)y_e), \quad \mathcal{D}(A_0) = V = \mathcal{D}(A^{\frac{1}{2}}),$

or equivalently, recalling (1.8),

(1.10b) $\qquad (A_0 y, z) = b(y_e, y, z) + b(y, y_e, z), \quad \forall\, y \in V,\ z \in H.$

The operator A_0 in (1.10) is well-defined $H \supset V = \mathcal{D}(A_0) \to H$. This follows from the estimate

(1.10c) $\qquad |A_0 y| \leq C_1 \|y_e\|_2 \|y\|, \quad \forall\, y \in V = \mathcal{D}(A_0) = \mathcal{D}(A^{\frac{1}{2}}),$

which is obtained directly by use of the definition (1.10a).

We have already noted below (1.6) that the operator $-\nu_0 A$ ($\nu_0 > 0$, the viscosity coefficient) is negative self-adjoint and has compact resolvent on H. Thus, $-\nu_0 A$ generates a s.c. analytic (self-adjoint) C_0-semigroup on H. It then follows from here and from $\mathcal{D}(A_0) = V = \mathcal{D}(A^{\frac{1}{2}})$, as noted in (1.6), that: *the perturbed operator*

(1.11) $\qquad \mathcal{A} = -(\nu_0 A + A_0),\ \text{with domain}\ \mathcal{D}(\mathcal{A}) = \mathcal{D}(A) = (H^2(\Omega))^d \cap V$

likewise has compact resolvent and generates a s.c. analytic C_0-semigroup on H. It follows from the above claim that the operator $\mathcal{A}$ has a finite number N of eigenvalues λ_j with Re $\lambda_j \geq 0$ (the unstable eigenvalues):

(1.12) $\qquad \text{Re}\,\lambda_{N+1} < 0 \leq \text{Re}\,\lambda_N \leq \cdots \leq \text{Re}\,\lambda_1.$

The eigenvalues are repeated according to their algebraic multiplicity ℓ_j. Let $\{\varphi_j\}_{j=1}^N$ be a corresponding system of *generalized eigenfunctions*, $\varphi_j = \varphi_j^1 + i\varphi_j^2$, $j = 1, \ldots, N$ of $\mathcal{A}$. (See [**K.1**, p. 41, 181].) More precisely, we shall denote by M the number of *distinct* unstable eigenvalues of $\mathcal{A}$, so that $\ell_1 + \ell_2 + \cdots + \ell_M = N$. We note at the outset that throughout this paper we shall denote with the same symbol $\mathcal{A}$ the *extension*, by transposition, $\mathcal{A}: H \to [\mathcal{D}(\mathcal{A}^*)]'$, duality with respect to H as a pivot space, of the original operator in (1.11). See (2.2), (2.4), (3.1.4), etc.

Domains of fractional powers and identification with Sobolev spaces. As already noted below (1.6), the fractional powers A^θ, $0 < \theta < 1$, are well-defined on H. Similarly, since $\mathcal{A}$ is the generator of a s.c. analytic semigroup, we may assume—without loss of generality (modulo an innocuous translation) for the list of relations below—that the fractional power $(-\mathcal{A})^\theta$ are likewise well-defined in H [**P.1**, p. 69], for otherwise we will have to take the fractional powers $(cI - \mathcal{A})^\theta$, for a constant $c > 0$ sufficiently large. We list below several identification relations to be frequently invoked in the sequel [**Tr.1**], [**L-M.1**], [**L-T.2**],

(1.13) $\quad \begin{aligned} \mathcal{D}\left((-\mathcal{A})^\theta\right) &= [\mathcal{D}(\mathcal{A}), H]_{1-\theta} = [\mathcal{D}(A), H]_{1-\theta} = \mathcal{D}(A^\theta) \\ &= [\mathcal{D}(\mathcal{A}^*), H]_{1-\theta} = \mathcal{D}((-\mathcal{A}^*)^\theta), \quad 0 < \theta < 1. \end{aligned}$

In particular, we shall repeatedly invoke the following cases:

(1.14) $\qquad \mathcal{D}\left(A^{\frac{1}{4}-\epsilon}\right) = \mathcal{D}\left((-\mathcal{A})^{\frac{1}{4}-\epsilon}\right) = \left(H^{\frac{1}{2}-2\epsilon}(\Omega)\right)^d \cap H,\ \epsilon > 0;$

(1.15) $$\mathcal{D}\left(\mathcal{A}^{\frac{1}{4}+\epsilon}\right) = \mathcal{D}\left((-\mathcal{A})^{\frac{1}{4}+\epsilon}\right) = \left(H_0^{\frac{1}{2}+2\epsilon}(\Omega)\right)^d \cap H,\ 0 < \epsilon < \frac{1}{2};$$

(1.16) $$\mathcal{D}\left(\mathcal{A}^{\frac{3}{4}-\epsilon}\right) = \mathcal{D}\left((-\mathcal{A})^{\frac{3}{4}-\epsilon}\right) = \left(H^{\frac{3}{2}-2\epsilon}(\Omega)\right)^d \cap V,\ 0 < \epsilon < \frac{1}{2};$$

(1.17) $$\mathcal{D}\left(\mathcal{A}^{\frac{3}{4}+\epsilon}\right) = \mathcal{D}\left((-\mathcal{A})^{\frac{3}{4}+\epsilon}\right) = \left(H^{\frac{3}{2}+2\epsilon}(\Omega)\right)^d \cap V \subset \left(H^{\frac{3}{2}+2\epsilon}(\Omega)\right)^d \cap H,$$
$$\epsilon \geq 0,$$

which can be obtained by interpolation between $\mathcal{D}(A)$, $\mathcal{D}(A^{\frac{1}{2}})$ and H [**W.1**], [**Tr.1**, p. 103], since $(\mathcal{A} - cI)$ is a positive operator [**Tr.1**, p. 91].

REMARK 1.1. Let $\{\varphi_j\}_{j=1}^\infty$ be the totality of the generalized eigenvectors of the operator $\mathcal{A}$ in (1.11). The following result holds true: the closure of the span of the generalized eigenvectors φ_j is all of H:

(1.18) $$\overline{\text{span}}\{\varphi_j\}_{j=1}^\infty = H.$$

PROOF. To establish (1.18), we apply Keldys' theorem in the simplified version given in [**D-S.1**, Vol. III, Theorem at p. 2374]. (See also its more general version in [**G-K.1**, Theorem 10.1, p. 276] or [**Tr.1**, Thm. 3, p. 394].) To this end, w.l.o.g. we take $\nu_0 = 1$ and rewrite $\mathcal{A}$ in (1.11) as

(1.19) $$-\mathcal{A} = A + A_0 = (I + K)A,\ \text{where}\ K = A_0 A^{-1}.$$

We now verify the assumptions of [**D-S.1**, Vol. III, Theorem at p. 2374]: (i) $A : H \supset \mathcal{D}(A) \to H$ is an unbounded, self-adjoint operator on the Hilbert space H; (ii) $A^{-1} \in \mathcal{L}(H)$, and, moreover, A^{-m} is a Hilbert-Schmidt operator for a sufficiently large m (this follows readily by using the well-known asymptotic behavior of the eigenvalues of A [**Tr.1**, p. 395]); (iii) the operator K in (1.19) is compact on H: in fact, $K = (A_0 A^{-\frac{1}{2}})A^{-\frac{1}{2}}$, where $A^{-\frac{1}{2}}$ is compact (see below (1.6)) while $(A_0 A^{-\frac{1}{2}})$ is a bounded operator as $\mathcal{D}(A_0) = \mathcal{D}(A^{\frac{1}{2}})$, via the closed graph theorem. Thus, Keldy's theorem applies and yields conclusion (1.18). We shall not use the result in (1.18) in this paper, however. □

CHAPTER 2

Main results

The following assumptions will be in effect throughout the paper.

Assumptions. (i) The boundary $\partial\Omega$ of Ω is a finite union of $d-1$ dimensional C^2-connected manifolds.

(ii) The steady-state solution (y_e, p_e) defined in (1.2) belongs to $((H^2(\Omega))^d \cap V) \times H^1(\Omega)$. [For $d = 2, 3$, this property is guaranteed by [**C-F.1**, Theorem 7.3, p. 59] on y_e, for $f_e \in H$, followed by [**C-F.1**, Theorem 3.11, p. 30] on p_e, for sufficiently smooth $\partial\Omega$.]

Preliminaries. The translated non-linear N-S problem. By the substitutions $y \to y_e + y$, $p \to p_e + p$ and $u \to y_e|_\Gamma + u$ (where $y_e|_\Gamma = 0$ is the Dirichlet trace of y_e on $\Gamma \equiv \partial\Omega$), we are readily led via (1.1), (1.2) to study the boundary *null stabilization* of the equation

$$
\begin{cases}
y_t - \nu_0 \Delta y + (y \cdot \nabla)y + (y_e \cdot \nabla)y + (y \cdot \nabla)y_e = \nabla p & \text{in } Q; \quad \text{(2.1a)}\\
\nabla \cdot y = 0 & \text{in } Q; \quad \text{(2.1b)}\\
y = u & \text{on } \Sigma; \quad \text{(2.1c)}\\
y(x, 0) = y^0(x) = y_0(x) - y_e(x) & \text{in } \Omega. \quad \text{(2.1d)}
\end{cases}
$$

Abstract model of the N-S problem (2.1) projected on H. We shall see in Section 3.1 that, under the pointwise compatibility condition (c.c.) $u \cdot \nu = 0$ on Σ of (1.1e) (whereby then $Py_t = y_t$, see Appendix A.1), application of the Leray projection P on (2.1a-d) leads to a corresponding equation in H, without the pressure terms, whose abstract version can be written as

(2.2) $\quad y_t - \mathcal{A}y + \mathcal{B}y = -\mathcal{A}Du \in [\mathcal{D}(\mathcal{A}^*)]'$, $\quad y(0) = y^0 \in H$, $\quad u \cdot \nu \equiv 0$ on Σ.

Here, the operator $\mathcal{A}$ in (2.2) is actually the *extension*, by transposition, $\mathcal{A} : H \to [\mathcal{D}(\mathcal{A}^*)]'$, duality with respect to H as a pivot space, of the original operator $\mathcal{A} : \mathcal{D}(\mathcal{A}) \to H$, defined in (1.11). Following established literature, we use the same symbol $\mathcal{A}$ for the original operator $\mathcal{A}$ in (1.11), as well as for its extension. Moreover, the non-linear operator $\mathcal{B}$ is defined in (1.7). See also Appendix A.2 complementing Section 3.1. Moreover, the operator $D : (L^2(\Gamma))^d \to (H^{\frac{1}{2}}(\Omega))^d \cap H \in \mathcal{D}(\mathcal{A}^{\frac{1}{4}-\epsilon})$ is defined in (3.1.3) below.

The translated linearized problem. PDE version. The translated linearized problem corresponding to (2.1) is then

(2.3a) $\quad\quad\quad\begin{cases} y_t - \nu_0 \Delta y + (y_e \cdot \nabla)y + (y \cdot \nabla)y_e = \nabla p & \text{in } Q; \\ \nabla \cdot y = 0 & \text{in } Q; \\ y = u & \text{on } \Sigma; \\ y(x,0) = y^0(x) & \text{in } \Omega. \end{cases}$

(2.3b)

(2.3c)

(2.3d)

Abstract model of problem (2.3) projected on H. Its *abstract* version on H is then (see (3.1.4a) and Appendix A)

(2.4) $\quad\quad y_t = \mathcal{A}y - \mathcal{A}Du \in [\mathcal{D}(\mathcal{A}^*)]', \quad y(0) = y^0 \in H; \quad u \cdot \nu \equiv 0.$

As in the case of Eqn. (2.2), the operator $\mathcal{A}$ in (2.4) is actually the *extension*, by transposition, $\mathcal{A} : H \to [\mathcal{D}(\mathcal{A}^*)]'$ of the original operator $\mathcal{A} : \mathcal{D}(\mathcal{A}) \to H$ defined in (1.11).

Main results: Case $d = 3$. The linearized model. We begin with the translated linearized problem (2.3) or its projected version (2.4). For the first result—the main result on problem (2.4)—essentially no assumptions are required.

THEOREM 2.1. *With reference to the linearized problem (2.3) or (2.4), the following results hold true:*

(i) Let $d = 3$ and assume further that Ω is connected. Then, given any $y^0 \in W \equiv (H^{\frac{1}{2}+\epsilon}(\Omega))^3 \cap H$, $\epsilon > 0$ arbitrary, there exists an open-loop, infinite-dimensional boundary control $u \in L^2(0, \infty; (L^2(\Gamma))^3)$, $u \cdot \nu \equiv 0$ on Σ, such that the corresponding solution y of (2.3) or (2.4) satisfies $y \in L^2(0, \infty; (H^{\frac{3}{2}+\epsilon}(\Omega))^3 \cap H) \cap H^{\frac{3}{4}+\frac{\epsilon}{2}}(0, \infty; H)$. Moreover, if y^0 vanishes on the portion Γ_0 of the boundary $\Gamma = \partial \Omega$, then u may be required to act only on the complementary part $\Gamma_1 = \Gamma \setminus \Gamma_0$ of the boundary. In particular, if y^0 vanishes on all of Γ, then u may be required to have an arbitrarily small support Γ_1, $\text{meas}(\Gamma_1) > 0$. [This is Theorem 3.5.1 along with Remark 3.5.1, illustrated by Figures 3.5.1 and 3.5.2.] The required boundary control is identified in (3.5.27).

(ii) Let $d = 3$. Then, the control u claimed in (i) cannot generally be finite-dimensional (as defined at the end of the Goal) except for a meager set of special initial conditions. [This is Proposition 3.1.3.]

The next result provides a more attractive and explicit open-loop boundary controller, but requires a Finite-Dimensional Spectral Assumption FDSA = (3.6.2). It is obtained by taking the 'real part version' of Proposition 3.7.1, that is, of the complexified system (3.4.1), where more details are provided explicitly.

THEOREM 2.2. *Let $d = 3$ and assume the FDSA = (3.6.2) (diagonalization of $\mathcal{A}$ restricted on its finite-dimensional unstable subspace). Let $\gamma_1 > 0$ be given arbitrary. Let N be defined by (1.12). Then:*

(a) there exist (infinite choices of) suitable boundary vectors $\{w_1, \ldots, w_K\}$, $K \leq N$, satisfying the rank conditions (3.6.16), $w_i \in \text{span}\{\partial_\nu \varphi_j^*\}_{j=1}^N$, φ_j^* being the eigenfunctions of the H-adjoint $\mathcal{A}^*$ corresponding to its unstable eigenvalues, so that $w_i \in (H^{\frac{1}{2}}(\Gamma))^3$; and there exist suitable vectors $\{u_1(t), u_2(t), \ldots, u_K(t)\} \subset$

2. MAIN RESULTS

$C([0,\infty); \mathbb{R}^K)$, $|u_i(t)| \leq c$, $t \geq 0$, and real vectors $\tilde{p}_1, \ldots, \tilde{p}_K$, such that the boundary open-loop control u:

(2.5)
$$u(t; y^0) - e^{-2\gamma_1 t}(y^0|_\Gamma) = e^{-\gamma_1 t}\sum_{i=1}^{2K} u_i(t)\tilde{w}_i - e^{-2\gamma_1 t}\sum_{i=1}^{K}(P_N y^0, \tilde{p}_i)_H(\tilde{w}_i - \tilde{w}_{i+K}),$$

$$\tilde{w}_i = Re\ w_i; \qquad \tilde{w}_{i+K} = -Im\ w_i, \qquad i = 1, \ldots, K,$$

obtained in (3.7.15) is stabilizing in the following sense: that is, once inserted in Eqn. (2.4), it causes the corresponding solution $y(t; y_0)$ to decay exponentially in H, with a decay rate γ_0, $0 < \gamma_0 < |Re\ \lambda_{N+1}|$, arbitrarily close to the absolute value of the real part of the first stable eigenvalue: there exists $C_{\gamma_0} > 0$ such that

(2.6)
$$|y(t; y^0)| \leq C_{\gamma_0} e^{-\gamma_0 t}|y^0|, \quad t \geq 0$$

[All of this is contained in Proposition 3.7.1 and Remark 3.7.1, after taking the real part version of the complexified system (3.4.1); Eqns. (3.7.15) yields (2.5).]

Finally, we may require the boundary vectors $\{w_i\}_{i=1}^{K}$ to have a (common) arbitrarily pre-assigned support Γ_1 on Γ, meas $\Gamma_1 > 0$. *[This is Remark 3.7.1.]*

In addition, if $y^0 \in (H^{\frac{1}{2}+\epsilon}(\Omega))^3 \cap H$, then $y \in L^2(0, \infty; (H^{\frac{3}{2}+\epsilon}(\Omega))^3 \cap H)$, see Proposition 3.7.1(ii), Eqns. (3.7.9a–c).

Case $d = 3$. Original N-S model (1.1). We now report the main result of the present paper, which provides the sought-after closed-loop boundary feedback control for the original N-S equations (1.1) [or its projected version (2.2)], which *exponentially stabilizes the stationary solution y_e of (1.1) in a neighborhood of y_e.* The stabilizing feedback control that we shall find is 'robust,' as it is expressed in terms of a Riccati operator R, which arises in an associated corresponding Optimal Control Problem (to be studied in Section 4). To state our (local) stabilizing result, we need to introduce the set

(2.7)
$$\mathcal{V}_\rho \equiv \left\{ y_0 \in W \equiv (H^{\frac{1}{2}+\epsilon}(\Omega))^3 \cap H : |y_0 - y_e|_W < \rho \right\}$$

of initial conditions y_0 of (1.1), whose distance in the norm of W from a stationary solution y_e is less than $\rho > 0$. Here, $\epsilon > 0$ arbitrary is fixed once and for all.

THEOREM 2.3 (Main Theorem). *Let $d = 3$ and assume further that Ω is connected. If $\rho > 0$ in (2.7) is sufficiently small, then: for each $y_0 \in \mathcal{V}_\rho$, there exists a unique fixed-point, mild, semigroup solution y (see Theorem 5.1) of the following closed-loop problem:*

(2.8a) $\quad \begin{cases} y_t(x,t) - \nu_0 \Delta y(x,t) + (y \cdot \nabla)y(x,t) = f_e(x) + \nabla p(x,t) & \text{in } G; \end{cases}$

(2.8b) $\quad \nabla \cdot y = 0 \qquad\qquad\qquad\qquad\qquad\qquad\qquad\qquad\qquad \text{in } G;$

(2.8c) $\quad y = \nu_0 \dfrac{\partial}{\partial \nu} R(y - y_e) \qquad\qquad\qquad\qquad\qquad\qquad \text{on } \Sigma;$

(2.8d) $\quad y(x,0) = y_0(x) \qquad\qquad\qquad\qquad\qquad\qquad\qquad \text{in } \Omega.$

obtained from (1.1) by replacing u with the boundary feedback control $u = \nu_0 \frac{\partial}{\partial \nu}R(y - y_e)$ having the following regularity and asymptotic properties:

(i)

(2.9)
$$(y - y_e) \in C([0, \infty); W) \cap L^2(0, \infty; (H^{\frac{3}{2}+\epsilon}(\Omega))^3 \cap H)$$

continuously in $y_0 \in W \equiv (H^{\frac{1}{2}+\epsilon}(\Omega))^3 \cap H$:

(2.10) $\quad |y(t) - y_e|_W^2 + \int_0^\infty |y(t) - y_e|_{(H^{\frac{3}{2}+\epsilon}(\Omega))^3 \cap H}^2 dt \leq C|y_0 - y_e|_W^2, \ t \geq 0.$

[This follows from Theorem 5.1 and Corollary 5.5, via the translation $y \to y_e + y$, etc., performed above problem (2.1).]

(ii) *there exist constants $M \geq 1$, $\omega > 0$ (independent of $\rho > 0$) such that such solution $y(t)$ satisfies*

(2.11) $\quad\quad\quad\quad |y(t) - y_e|_W \leq Me^{-\omega t}|y_0 - y_e|_W, \quad t \geq 0.$

[This follows from Theorem 6.1(i), via the translation $y \to y + y_e$, etc., performed above problem (2.1).]

Here R is a Riccati operator, in the sense that it [arises in the Optimal Control Problem of Section 4.1 and] satisfies the Algebraic Riccati Equation (4.5.1) below. The operator R is positive self-adjoint on H and, moreover, $R \in \mathcal{L}(W; W')$ where W' is the dual of W with respect to H as a pivot space. In addition [Proposition 4.1.4],

$$c|x|_W^2 \leq (Rx, x)_H \leq C|x|_W^2, \ 0 < c < C < \infty, \ \forall \ x \in W,$$

so that the $|R^{\frac{1}{2}}x|$-norm is equivalent to the W-norm.

REMARK 2.1. *In Theorem 2.3, if the I. C. $y_0 \in \mathcal{V}_\rho$ is, moreover, suitably smoother, then one obtains accordingly stronger results: (i) regularity results sharper than (2.9), (2.10), as well as (ii) stabilization results in a higher norm than the W-norm in (2.11)*

More precisely, assume that

(2.12) $\quad\quad\quad\quad y_0 \in \mathcal{V}_\rho \cap \mathcal{D}(A_R), \ \rho > 0 \text{ sufficiently small},$

where A_R is the infinitesimal generator, identified in (4.2.9). Then, the solution y of problem (2.8) claimed in Theorem 2.3 satisfies the following regularity and asymptotic properties:

(i)

(2.13a) $\quad y - y_e \ \in \ C([0, T]; (H^{\frac{3}{2}+\epsilon}(\Omega))^3 \cap H) \cap L^2(0, \infty; (H^{\frac{3}{2}+\epsilon}(\Omega))^3 \cap H);$

(2.13b) $\quad\quad\quad y_t \ \in \ C([0, T]; (H^{\frac{1}{2}+\epsilon}(\Omega))^3 \cap H) \cap L^2(0, \infty; (H^{\frac{3}{2}+\epsilon}(\Omega))^3 \cap H),$

i.e., setting $W \equiv (H^{\frac{1}{2}+\epsilon}(\Omega))^3 \cap H; \ Z \equiv (H^{\frac{3}{2}+\epsilon}(\Omega))^3 \cap H,$

(2.14) $\quad |y(t) - y_e|_Z^2 + |y_t(t)|_W^2 + \int_0^\infty [|y(t) - y_e|_Z^2 + |y_t(t)|_Z^2] dt \leq C|y_0 - y_e|_W^2,$

for some constant C. [This follows from Theorem 5.2, via the translation $y \to y + y_e$, etc., performed above problem (2.1).]

(ii) There exist constants $M_1 \geq 1$, $\omega_1 > 0$ (independent of $\rho > 0$), such that

(2.15) $\quad\quad\quad\quad |y(t) - y_e|_Z^2 \leq M_1 e^{-\omega_1 t}|y_0 - y_e|_Z, \quad t \geq 0.$

[This follows from Theorem 6.1(ii) via the translation $y \to y + y_e$, etc.] With reference to (2.12), we also recall that $\mathcal{D}(A_R) \subset Z \equiv (H^{\frac{3}{2}+\epsilon}(\Omega))^3 \cap H$ (see (4.2.10)) and that the s.c. semigroup $S(t) \equiv e^{A_R t}$ on W, see (4.2.3), defines the optimal solution of the Optimal Control Problem studied in Section 4.

2. MAIN RESULTS

Case $d = 2$. The linearized model.

THEOREM 2.4. *Let $d = 2$. With reference to the linearized problem (2.4), the following result holds true: Given $y^0 \in (H^{\frac{1}{2}-\epsilon}(\Omega))^2 \cap H$, $\epsilon > 0$ arbitrary, there exists an open-loop, infinite-dimensional boundary control $u \in L^2(0, \infty; (L^2(\Gamma_1))^2)$, $u \equiv 0$ on $\Gamma \setminus \Gamma_1$, $u \cdot \nu \equiv 0$ on Σ, where Γ_1 is an arbitrarily pre-assigned part of the boundary Γ, with meas $\Gamma_1 > 0$, such that the corresponding solutions y of (2.4) satisfies $y \in L^2(0, \infty; (H^{\frac{3}{2}-\epsilon}(\Omega))^2 \cap H)$. [This is Theorem B.1.1, Appendix B.]*

Under the additional FDSA = (3.6.2) [diagonalization of $\mathcal{A}$ restricted on its finite-dimensional unstable subspace], a corresponding *closed-loop boundary feedback stabilization* result of the linearized system (2.4), $d = 2, 3$, with *finite-dimensional* feedback controllers acting on an *arbitrarily small* portion Γ_1 of the boundary $\Gamma = \partial\Omega$, meas $\Gamma_1 > 0$, is given in Proposition B.2.1, Appendix B, to which we refer for a precise statement and its proof: in it, $y^0 \in (H^{\frac{1}{2}-\epsilon}(\Omega))^d \cap H$ and the corresponding solution y lies in $L^2(0, \infty; (H^{\frac{3}{2}-\epsilon}(\Omega))^d \cap H)$.

Here, instead, we concentrate on the full N-S model (1.1), for $d = 2$.

Case $d = 2$. Original N-S model (1.1). The next Theorem 2.5 follows from Theorem 2.4 the same way that Theorem 2.3 follows from Theorem 2.1. In this case, the Riccati theory of the literature [**L-T.1**, Section 6.1, p. 51], [**L-T.2**, Vol. 1, Theorem 3.2.1(ii), p. 187]—slightly refined, as described in Appendices B.3 and B.4 below—replaces the more demanding and less satisfying theory of Section 4, see Appendices B.3 and B.4. In particular, the corresponding Riccati operator, $\tilde{R}$ in (2.17) below, satisfies the standard Algebraic Riccati Equation, even on H this time, which originates from the OCP with cost $\{|y|^2_{L^2(0,\infty;H)} + |u|^2_{L^2(0,\infty;U)}\}$ (as in [**L-T.1**, Eqn. (6.2), p. 51], [**L-T.2**, Vol. 1, Eqn. (3.1.2), p. 180]), instead of the cost (4.1.3) for $d = 3$.

THEOREM 2.5. *Let $d = 2$. Then, the same statement as that of Theorem 2.3 for $d = 3$ holds true, once the following changes are made:*

(i) space of initial data in (2.7):

(2.16a) $\qquad W \equiv \left(H^{\frac{1}{2}+\epsilon}(\Omega)\right)^3 \cap H$ *is replaced by* $\tilde{W} \equiv \left(H^{\frac{1}{2}-\epsilon}(\Omega)\right)^2 \cap H$;

so that $\mathcal{V}_\rho$ in (2.7) is now replaced by

(2.16b) $\qquad\qquad\qquad \tilde{\mathcal{V}}_\rho = \{y_0 \in \tilde{W} : |y_0 - y_e|_{\tilde{W}} < \rho\};$

(ii) feedback control in (2.8c)

(2.17)

$$y = \nu_0 \frac{\partial}{\partial \nu} \tilde{R}(y - y_e) \text{ on } \Sigma \text{ is replaced by } \begin{cases} y = \nu_0 \frac{\partial}{\partial \nu} \tilde{R}(y - y_e) & \text{on } \Sigma_1; \\ y \equiv 0 & \text{on } \Sigma_2, \end{cases}$$

where Γ_1, meas $\Gamma_1 > 0$, is arbitrarily pre-assigned, while $\Gamma_2 = \Gamma \setminus \Gamma_1$: here $\tilde{R}$ is the Riccati operator identified just above the theorem's statement, that is, satisfying the ARE: $\mathcal{A}^*\tilde{R} + \tilde{R}\mathcal{A} + I = (B^*\tilde{R})^*(B^*\tilde{R})$, $B = -\mathcal{A}D$ in (2.4) (a marked improvement over Eqn. (4.5.1) for $d = 3$): see Appendices B.3 and B.4;

(iii) space of solutions y to problem (2.8):

(2.18) $\qquad\qquad \left(H^{\frac{3}{2}+\epsilon}(\Omega)\right)^3 \cap H$ *is replaced by* $\left(H^{\frac{3}{2}-\epsilon}(\Omega)\right)^2 \cap H.$

THEOREM 2.6. *Let $d = 2$ and assume the FDSA = (3.6.2), as stated in Theorem 2.2. There exist boundary functions $\psi_i \in (H^{\frac{1}{2}}(\Gamma_1))^2$:*

$$\{\psi_i\}_{i=1}^{K} \subset X_N^1 \equiv \text{span}\left\{\partial_\nu \varphi_j^{*1}\Big|_{\Gamma_1}\right\}_{j=1}^{N}; \quad (2.19)$$

$$\{\psi_i\}_{i=K+1}^{2K} \subset X_N^2 \equiv \text{span}\left\{\partial_\nu \varphi_j^{*2}\Big|_{\Gamma_1}\right\}_{j=1}^{N}, \quad (2.20)$$

Γ_1 *being an arbitrarily preassigned part of* $\Gamma = \partial \Omega$, *meas* $\Gamma_1 > 0$,

$$\varphi_j^* = \varphi_j^{*1} + i\varphi_j^{*2} = \text{eigenfunction of } \mathcal{A}^*,$$

and there exists a linear operator $\Phi = [\Phi_1, \ldots, \Phi_{2K}] \in \mathcal{L}(\tilde{W}; \mathbb{R}^{2K})$; $\tilde{W} \equiv (H^{\frac{1}{2}-\epsilon}(\Omega))^2 \cap H$, *such that the feedback controller*

$$u(x,t) = \sum_{i=1}^{2K} \Phi_i(y - y_e)(t)\psi_i(x), \quad x \in \Gamma_1, \ t \geq 0, \quad (2.21)$$

once inserted into Eqn. (1.1c) of the N-S model (1.1), exponentially stabilizes the steady-state solution y_e to (1.1) in a neighborhood $\tilde{\mathcal{V}}_\rho$ of y_e as in (2.16b), for suitable $\rho > 0$. More precisely, if $\rho > 0$ is sufficiently small, then for each $y_0 \in \tilde{\mathcal{V}}_\rho$, there exists a fixed-point, mild semigroup solution y,

$$y \in C([0,\infty); \tilde{W}) \cap L^2\left(0,\infty; (H^{\frac{3}{2}-\epsilon}(\Omega))^2 \cap H\right), \quad (2.22)$$

of the closed-loop system, where $\Gamma_2 = \Gamma \setminus \Gamma_1$:

$$\begin{cases} y_t - \nu_0 \Delta y + (y \cdot \nabla)y = f_e + \nabla p & \text{in } G; \quad (2.23\text{a}) \\ \nabla \cdot y = 0 & \text{in } G; \quad (2.23\text{b}) \\ y = \sum_{i=1}^{2K} \Phi_i(y - y_e)(t)\psi_i(x) & \text{in } \Sigma_1; \quad (2.23\text{c}) \\ y = 0 & \text{in } \Sigma_2; \quad (2.23\text{d}) \\ y(x,0) = y_0(x) & \text{in } \Omega, \quad (2.23\text{e}) \end{cases}$$

such that the counterpart properties of (2.9)–(2.15) hold true:

$$(y - y_e) \in C([0,\infty); \tilde{W}) \cap L^2(0,\infty; \tilde{Z}) \quad (2.24)$$

continuously in $y_0 \in \tilde{W} \equiv (H^{\frac{1}{2}-\epsilon}(\Omega))^2 \cap H$, where $\tilde{Z} \equiv (H^{\frac{3}{2}-\epsilon}(\Omega))^2 \cap H$. Moreover, there exist $M \geq 1$, $\omega > 0$ (independent of $\rho > 0$), such that

$$|y(t) - y_e|_{\tilde{W}} \leq M e^{-\omega t} |y_0 - y_e|_{\tilde{W}}, \quad t \geq 0. \quad (2.25)$$

The effective (non-trivial) number of the $2K$-controllers is $\leq N$.

The proofs of Theorem 2.5 and Theorem 2.6 are indicated in Appendices B.3, B.4. The first (respectively, the second) follows from Theorem 2.4, which is Theorem B.1.1 in Appendix B (respectively, Theorem B.2.1 in Appendix B) by using the same approach that yields Theorem 2.3 from Theorem 2.1 for $d = 3$, albeit in a much more simplified setting, due to the more amenable topological setting for $d = 2$ over $d = 3$. In particular, for $d = 2$, the treatment of Section 4 on the Optimal Control Problem and related Algebraic Riccati Equation can be dispensed with altogether, and the available literature on this subject as in [**L-T.1**], [**L-T.2**, Vol. 1] can be simply invoked instead.

CHAPTER 3

Proof of Theorems 2.1 and 2.2 on the linearized system (2.4): $d = 3$

3.1. Abstract models of the linearized problem (2.3). Regularity

The present Section 3.1 is to be integrated with technical material collected in Appendix A, to which we shall make reference.

The linearized dynamical system. First, for convenience, we denote by $\mathbf{A}$ the differential operator in (2.3a)

(3.1.1) $$\mathbf{A}y = -\nu_0 \Delta y + (y_e \cdot \nabla)y + (y \cdot \nabla)y_e.$$

In this notation, we rewrite the linearized problem (2.3) as

$$\begin{cases}
y_t + \mathbf{A}y = \nabla p & \text{in } \Omega \times (0, \infty); \quad \text{(3.1.2a)} \\
\nabla \cdot y = 0 & \text{in } \Omega; \quad \text{(3.1.2b)} \\
y(x, 0) = y^0(x) & \text{in } \Omega; \quad \text{(3.1.2c)} \\
y = g & \text{in } \partial\Omega \times (0, \infty) \equiv \Sigma, \quad \text{(3.1.2d)}
\end{cases}$$

where g is the boundary control (of suitable regularity, say $g \in L_2(0, T; (H^s(\Gamma)^d))$, $s \geq 0$; in particular, with $s = \frac{1}{2}$, as it will be *a-fortiori* the case in our study below) satisfying the boundary compatibility condition

either pointwise: or at least in integral form:

(3.1.2e) $\quad g \cdot \nu \equiv 0$ on Σ; $\qquad \int_\Gamma g \cdot \nu d\Gamma = 0, \ t > 0,$

as in (1.1e), to be specified on a case-by-case basis.

The corresponding stationary problem. In view of (1.5a–b), the process of applying the Leray orthogonal projector P—to be carried out below—eliminates the pressure from the equations [**C-F.1**, p. 47], when acting on $(L^2(\Omega))^d$. This is integrated by Section A.1 of Appendix A dealing with the Leray projector outside the space $(L^2(\Omega))^d$. Accordingly, we introduce the stationary problem

(3.1.3a)
(3.1.3b) $\quad \psi \equiv Dg \in \left(H^{s+\frac{1}{2}}(\Omega)\right)^d \iff \begin{cases} \mathbf{A}\psi = \nabla p^* & \text{in } \Omega; \\ \nabla \cdot \psi = 0 & \text{in } \Omega; \\ \psi = g \in (H^s(\Gamma))^d & \text{in } \Gamma, \end{cases}$
(3.1.3c)

with $s \geq -\frac{1}{2}$.

Orientation. More precisely, in Section A.2 of Appendix A, we consider the variation of problem (3.1.1) with the (perhaps necessary) translation $(k + \mathbb{A})$ in place of $\mathbb{A}$, for a suitable constant $k > 0$ sufficiently large. Then, Theorem A.2.1 of Section A.2 shows the existence of the Dirichlet map D_k, depending on k, with the required regularity properties (A.2.4b), for $s \geq \frac{1}{2}$. This result on D_k is then extended in Theorem A.2.2 to hold true for all $s \geq -\frac{1}{2}$: such extension is achieved by a duality argument on the independent Proposition 3.2.1, i.e., formula (A.2.6), which for $k = 0$ becomes formula (3.2.2). Finally, Remark A.2.1 of Section A.2 of Appendix A points out that, "generically," one can take $k = 0$, essentially with no loss of generality in the analysis. To streamline the exposition in the text, *we shall therefore take $k = 0$ and denote $D_{k=0}$ simply by D.*

As documented in Section A.2, and with the understanding of the above Orientation, we shall consider two cases:

Case 1: $g \cdot \nu \equiv 0$ on Σ in (3.1.2e). Here, given the boundary datum $g \in (H^s(\Gamma))^d$, $s \geq -\frac{1}{2}$, satisfying the pointwise c.c. $g \cdot \nu \equiv 0$ on Σ in (3.1.2e), there exists a unique solution $\psi \equiv Dg \in \left(H^{s+\frac{1}{2}}(\Omega)\right)^d \cap H$ (Theorems A.2.1 and A.2.2); that is,

$$(3.1.3d) \quad D : \text{continuous } (H^s(\Gamma))^d \longrightarrow \left(H^{s+\frac{1}{2}}(\Omega)\right)^d \cap H \text{ with } g \cdot \nu \equiv 0 \text{ on } \Gamma,$$

in particular for $s = 0$:

$$(3.1.3e) \quad D: \text{continuous } (L^2(\Gamma))^d \longrightarrow (H^{\frac{1}{2}}(\Omega))^d \cap H \subset (H^{\frac{1}{2}-2\epsilon}(\Omega))^d \cap H$$
$$\text{with } g \cdot \nu = 0 \text{ on } \Gamma, \quad = \mathcal{D}((-\mathcal{A})^{\frac{1}{4}-\epsilon})$$

where H is the basic space (1.5a) of the Navier-Stokes equations. Thus, recalling $H^\perp$ from (1.5b), we have that projecting Eqn. (3.1.2a) onto $H^\perp$ eliminates the pressure term. Moreover, returning to problem (3.1.2) with $g \cdot \nu \equiv 0$, we then have $y \in H$, hence $Py_t = y_t$ in this case, see also Section A.1 of Appendix A.

Case 2: $\int_\Gamma g \cdot \nu \, d\Gamma = 0$ in (3.1.2e). In order to assert a unique solution $g \in (H^s(\Gamma))^d \to \psi \equiv Dg \in (H^{s+\frac{1}{2}}(\Omega))^d$ [modulo a possible translation of $\mathcal{A}$], it would suffice to impose on g the weaker integral c.c. $\int_\Gamma g \cdot \nu \, d\Gamma = 0$ [**F-T.1**], [**Te.1**], [**G.1**]. However, in this case, we would have $Dg = \psi \notin H$, a serious disadvantage in the computations below ((i) e.g., returning to problem (3.1.2) we cannot now assert that $y \in H$, hence that $Py_t = y_t$; (ii) moreover, see Remark 3.2.1 for an illustration of the complications that arise in the computations to obtain $D^*\mathcal{A}^*$ in Section 3.2 below), and we need to apply the Leray projector to assert $PDg = P\psi \in H$. Hence, in this case, we have for $s \geq 0$:

$$(3.1.3f) \quad PD : \text{continuous } (H^s(\Gamma))^d \to (H^{s+\frac{1}{2}}(\Omega))^d \cap H \text{ with } \int_\Gamma g \cdot \nu \, d\Gamma = 0.$$

For most of our treatment, we shall fall in Case 1 above. Thus, it is the double requirement $y \in H$ and so $y_t = Py_t$, see also Section A.1 of Appendix A; $\psi = Dg \in H$ that has forced us to impose the much more restrictive condition $g \cdot \nu \equiv 0$ on Σ (3.1.2e). We shall see in Lemma 3.3.1 below a quite useful case for our treatment where this condition will be automatically satisfied.

However, in Section 3.5, we shall also encounter a situation (see problem (3.5.17) for $\hat{y}$) which intrinsically belongs to Case 2 above. To handle also this case, we shall provide the abstract model of problem (3.1.2a–d) for both cases contemplated in (3.1.2e).

3.1. ABSTRACT MODELS OF THE LINEARIZED PROBLEM (2.3). REGULARITY

The abstract model. We refer to Theorem A.2.1(ii), Eqn. (A.2.5) of Section A.2, Appendix A, as specialized to the case $k = 0$. In the setting of (3.1.3d), we have seen in the aforementioned theorem that (as anticipated by (2.4)), the abstract model for problem (3.1.2) is

(3.1.4a) $\qquad y' - \mathcal{A}y = -\mathcal{A}Dg \in [\mathcal{D}(\mathcal{A}^*)]', \quad \psi = Dg \in H,$

$\qquad\qquad \nabla \cdot \psi = 0 \text{ in } G; \quad g \cdot \nu = 0 \text{ on } \Sigma.$

(3.1.4b) $\qquad (Py)' - \mathcal{A}(Py) = -\mathcal{A}PDg \in [\mathcal{D}(\mathcal{A}^*)]', \quad P\psi = PDg \in H,$

$\qquad\qquad \nabla \cdot \psi = 0 \text{ in } G; \quad \int_\Gamma g \cdot \nu \, d\Gamma = 0, \; t \geq 0.$

Here $\mathcal{A}$ actually denotes the *extension*, by transposition, $\mathcal{A} : H \to [\mathcal{D}(\mathcal{A}^*)]'$, duality with respect to H as a pivot space, of the original operator $\mathcal{A} : \mathcal{D}(\mathcal{A}) \to H$ in (1.11). This was noted below (1.12), (2.2), and (2.4). This is obtained following the established procedure [**L-T.1**] long used in boundary control theory for PDEs, which is displayed in the general case ($k \neq 0$) in the proof of Theorem A.2.1(iii) in Section A.2 of Appendix A. For instance, to obtain (3.1.4a) for $k = 0$, we subtract $\mathbb{A}\psi = \mathbb{A}Dg = \nabla p^*$ from the dynamical equation (3.1.2a), thus obtaining by (3.1.3)

$$\begin{cases} (3.1.5\text{a}) & y_t + \mathbf{A}(y - \psi) = \nabla(p - p^*) & \text{in } \Omega \times (0, \infty); \\ (3.1.5\text{b}) & \nabla \cdot (y - \psi) = 0 & \text{in } \Omega; \\ (3.1.5\text{c}) & y(x,0) - \psi(x) = y^0(x) - \psi(x) & \text{in } \Omega; \\ (3.1.5\text{d}) & y - \psi = g - g = 0 & \text{in } \Gamma \times (0, \infty), \end{cases}$$

where $[y - \psi] \in H$. Next, we apply the Leray orthogonal projector to (3.1.5a) so that $\nabla(p - p^*)$ is eliminated, see (1.5a-b), while $Py_t = y_t$ since $y \in H$ (via Proposition A.1.1 in Section A.1, Appendix A) in this case where $g \cdot \nu = 0$. We then recall the definitions of A, A_0 in (1.6), (1.10) to obtain the abstract equation $y' + (\nu_0 A + A_0)(y - \psi) = 0$, whereby then (3.1.4a) follows recalling $\mathcal{A}$ from (1.11). Instead, in the case of (3.1.4b), we have from (3.1.5) that $P(y - \psi) = (y - \psi) \in H$. Thus applying P to $y_t + \mathbb{A}P(y - \psi) = \nabla(p - p^*)$ [which therefore is the same as (3.1.5a)], we obtain (3.1.4b).

Regularity of (3.1.4) for g non-smooth. Model (3.1.4) is suitable for a non-regular boundary datum, say $g \in L_2(0, T; (L^2(\Gamma))^d)$. The solution of model (3.1.4) is given by the variation of parameter formula

(3.1.6a) $\qquad (Py)(t) = e^{\mathcal{A}t}Py_0 + (Lg)(t);$

(3.1.6b) $\qquad (Lg)(t) = -\mathcal{A}\int_0^t e^{\mathcal{A}(t-\tau)}PDg(\tau)d\tau$

$\qquad\qquad = \int_0^t (cI - \mathcal{A})^{\frac{3}{4}+\epsilon} e^{\mathcal{A}(t-\tau)} (cI - \mathcal{A})^{\frac{1}{4}-\epsilon} PDg(\tau)d\tau$

$\qquad\qquad - c\int_0^t e^{\mathcal{A}(t-\tau)} PDg(\tau)d\tau,$

in suitable function spaces, e.g., specified below, where the Leray projector P is redundant on each of the terms in (3.1.6) ($Py = y$; $PDg = Dg$) and hence omitted, if $g \cdot \nu \equiv 0$ in Σ, and where we have recalled (3.1.3e). Moreover, as the regularity results of the present section are given on a *finite* interval $[0, T]$, we may then assume w.l.o.g. that the fractional powers of $(-\mathcal{A})$ are well-defined and streamline (3.1.6b) by setting $c = 0$, thus eliminating the last, benign integral term; see the last part of Section 1, Eqns. (1.14)–(1.17).

LEMMA 3.1.1. *With reference to (3.1.6), or (3.1.4) [orthogonal projection of problem (3.1.2) on the space H], we have:*

(a)

$$(3.1.7) \qquad y^0 \in \left[\mathcal{D}((-\mathcal{A})^{\frac{1}{4}+\epsilon})\right]', \ equivalently, \ (-\mathcal{A})^{-(\frac{1}{4}+\epsilon)}y^0 \in H,$$

$$(3.1.8) \quad \Rightarrow e^{\mathcal{A}t}y^0 = (-\mathcal{A})^{\frac{1}{4}+\epsilon}e^{\mathcal{A}t}(-\mathcal{A})^{-(\frac{1}{4}+\epsilon)}y^0$$

$$\in C\left([0,T]; \left[\mathcal{D}((-\mathcal{A})^{\frac{1}{4}+\epsilon})\right]'\right) \cap L^2\left(0,T; \mathcal{D}((-\mathcal{A})^{\frac{1}{4}-\epsilon})\right)$$

continuously, where $\mathcal{D}((-\mathcal{A})^{\frac{1}{4}-\epsilon}) = (H^{\frac{1}{2}-2\epsilon}(\Omega))^d \cap H$ *by (1.14);*

(b)

$$(3.1.9) \qquad g \in L^2(0,T; (L^2(\Gamma))^d), \quad \int_\Gamma g \cdot \nu d\Gamma \equiv 0 \ on \ \Gamma, \ a.e. \ in \ t;$$

$$(3.1.10) \quad \Rightarrow Lg \in C\left([0,T]; \left[\mathcal{D}((-\mathcal{A})^{\frac{1}{4}+\epsilon})\right]'\right) \cap L^2\left(0,T; \mathcal{D}((-\mathcal{A})^{\frac{1}{4}-\epsilon})\right)$$

continuously.

(c) *Under the above assumptions on y^0 and g in (3.1.7), (3.1.9), we have*

$$(3.1.11) \qquad y \in C\left([0,T]; \left[\mathcal{D}((-\mathcal{A})^{\frac{1}{4}+\epsilon})\right]'\right) \cap L^2\left(0,T; \mathcal{D}((-\mathcal{A})^{\frac{1}{4}-\epsilon})\right).$$

PROOF. See, e.g., [**B-T.1**, Appendix], and (3.1.3e). □

The abstract model revisited, for g smoother. When g is smoother, in particular if $g_t \in L^2(0,T; (L^2(\Gamma))^d)$, and still, of course, $g \cdot \nu \equiv 0$, whereby then $y \in H$ and $Dg \in H$, a more suitable abstract model of problem (3.1.2) is

$$(3.1.12) \qquad \eta_t = \mathcal{A}\eta - Dg_t; \quad \eta(t) = y(t) - Dg(t); \quad t \geq 0, \ g \cdot \nu \equiv 0 \ \text{on} \ \Sigma,$$

whose variation of parameter solution is:

$$(3.1.13a) \qquad \eta(t) = e^{\mathcal{A}t}[y^0 - Dg(0)] - (\mathcal{K}Dg_t)(t)$$

$$(3.1.13b) \qquad (\mathcal{K}f)(t) = \int_0^t e^{\mathcal{A}(t-\tau)}f(\tau)d\tau.$$

This can be seen, as usual [**L-T.1**], [**L-T.2**], as follows: we return to problem (3.1.2) and obtain, with $\psi = Dg$ defined by (3.1.3) under (3.1.2e):

$$(3.1.14) \qquad (y - Dg)_t + \mathbf{A}(y - Dg) = -Dg_t + \nabla(p - p^*) \ \text{in} \ \Omega \times (0, \infty),$$

instead of (3.1.5a), while Eqns. (3.1.2b-c-d) continue to hold true. Setting $\eta = y - Dg \in H$ and applying the Leray projector P on (3.1.14) onto H, see also Section A.1 of Appendix A, results in (3.1.12), as desired, as $P\eta_t = \eta_t$, $PDg_t = Dg_t$.

Regularity of (3.1.12). In subsequent sections (Section 3.7; Appendix B), we shall fall into the following setting.

(3.1.15) $y^0 \in \mathcal{D}((-\mathcal{A})^{\frac{1}{4}-\epsilon}) \equiv (H^{\frac{1}{2}-2\epsilon}(\Omega))^d \cap H; \quad g \in H^1\left([0,T]; (H^{\frac{1}{2}}(\Gamma))^d\right);$

hence $Dg \in H^1([0,T]; (H^1(\Omega)^d \cap \Pi)$.

[Indeed, g will be much smoother in time, at least, $g \in C^n([0,T];(H^{\frac{1}{2}}(\Gamma))^d)$ for any $n = 1, 2, \ldots$.]

LEMMA 3.1.2. *With reference to (3.1.12), in particular with $g \cdot \nu \equiv 0$, we have:*
(i) Assume the regularity of y^0, g in (3.1.15). Then:

(3.1.16) $\quad \eta, y \in C\left([0,T]; \mathcal{D}((-\mathcal{A})^{\frac{1}{4}-\epsilon})\right) \cap L^2\left(0,T; \mathcal{D}((-\mathcal{A})^{\frac{3}{4}-\epsilon})\right)$

continuously, where $\mathcal{D}((-\mathcal{A})^{\frac{1}{4}-\epsilon}) = (H^{\frac{1}{2}-2\epsilon}(\Omega))^d \cap H; \mathcal{D}((-\mathcal{A})^{\frac{3}{4}-\epsilon}) = H^{\frac{3}{2}-2\epsilon}(\Omega) \cap H$, by (1.14), (1.16).
(ii) In addition, assume now the compatibility condition, see (1.15):

(3.1.17) $\quad [y^0 - Dg(0)] \in \mathcal{D}((-\mathcal{A})^{\frac{1}{4}+\epsilon}) = (H_0^{\frac{1}{2}+2\epsilon}(\Omega))^d \cap H$, *hence $y^0|_\Gamma = g(0)$.*

Then:

(3.1.18a) $\quad \begin{cases} \eta \in C\left([0,T]; \mathcal{D}\left((-\mathcal{A})^{\frac{1}{4}+\epsilon}\right) \equiv \left(H_0^{\frac{1}{2}+2\epsilon}(\Omega)\right)^d \cap H\right) \\ \qquad \cap L^2\left(0,T; \mathcal{D}\left((-\mathcal{A})^{\frac{3}{4}+\epsilon}\right) \equiv \left(H^{\frac{3}{2}+2\epsilon}(\Omega)\right)^d \cap V\right); \end{cases}$

(3.1.18b) $\quad y \in C\left([0,T]; \left(H^{\frac{1}{2}+2\epsilon}(\Omega)\right)^d \cap H\right) \cap L^2\left(0,T; \left(H^{\frac{3}{2}+2\epsilon}(\Omega)\right)^d \cap H\right);$

(3.1.18c) $\quad y_t = -P\mathbb{A}y \in L^2\left(0,T; \left(H^{-\frac{1}{2}+2\epsilon}(\Omega)\right)^d\right);$

(3.1.18d) $\quad y|_\Gamma = g \in C\left([0,T]; (H^{2\epsilon}(\Gamma))^d\right).$

PROOF. (i) Under present assumptions in (3.1.15), we have $y^0 - Dg(0) \in \mathcal{D}((-\mathcal{A})^{\frac{1}{4}-\epsilon})$, by (3.1.3e), and hence

(3.1.19) $\quad e^{\mathcal{A}t}[y^0 - Dg(0)] \in C\left([0,T]; \mathcal{D}((-\mathcal{A})^{\frac{1}{4}-\epsilon})\right) \cap L^2\left(0,T; \mathcal{D}((-\mathcal{A})^{\frac{3}{4}-\epsilon})\right).$

Moreover, with $(-\mathcal{A})^{\frac{1}{4}-\epsilon} Dg_t \in L^2(0,T;H)$, *a-fortiori* by assumption (3.1.15) we have

(3.1.20) $\quad (\mathcal{K}Dg_t)(t) = \int_0^t (-\mathcal{A})^{-\frac{1}{4}+\epsilon} e^{\mathcal{A}(t-\tau)}(-\mathcal{A})^{\frac{1}{4}-\epsilon} Dg_t(\tau) d\tau$

$\in C\left([0,T]; \mathcal{D}((-\mathcal{A})^{\frac{3}{4}-\epsilon})\right) \cap L^2\left(0,T; \mathcal{D}((-\mathcal{A})^{\frac{5}{4}-\epsilon})\right),$

and (3.1.16) for η follows from (3.1.13a), (3.1.19), (3.1.20). Then $y(t) = \eta(t) + Dg(t)$, with $Dg \in C\left([0,T]; (H^1(\Omega))^d \cap H\right)$, yields (3.1.16) also for y.

(ii) Now, under (3.1.17), we improve (3.1.19) to

(3.1.21) $\quad e^{\mathcal{A}t}[y^0 - Dg(0)] \in C\left([0,T]; \mathcal{D}((-\mathcal{A})^{\frac{1}{4}+\epsilon})\right) \cap L^2\left(0,T; \mathcal{D}((-\mathcal{A})^{\frac{3}{4}+\epsilon})\right),$

while (3.1.20) continues to hold true. This time, therefore, via (3.1.13), we obtain (3.1.18a) for η: then (3.1.18b) follows from $y = \eta + Dg$; while (3.1.18c) follows from the equation: $y_t + P\mathbb{A}y = 0$ in $\Omega \times (0,\infty)$, $Py_t = y_t$. Then (3.1.18d) follows by trace theory applied on (3.1.18b). [Due to the boundary regularity of $y|_\Gamma = g$, then the c.c. $y^0|_\Gamma = g(0)$ is an immediate consequence.] □

Consequences. For $d = 3$, *boundary stabilization of (the linearized system (2.4) in $(H^{\frac{3}{2}+2\epsilon}(\Omega))^d \cap H$, hence of) the Navier-Stokes problem (1.1) with finite-dimensional controller is not possible* (except for a meager set of very special initial conditions).

The content of Lemma 3.1.2 is very enlightening for our development to follow. In fact, it turns out that, in order to obtain the sought-after solution of the local boundary stabilization problem of the original Navier-Stokes model (1.1), for $d = 3$, the non-linear term $(y \cdot \nabla)y$ in (1.1) imposes—via a Sobolev embedding and multiplier arguments for $d = 3$, see Eqn. (5.18b) below—that the following auxiliary problem be preliminarily resolved for the *corresponding linearized model* (3.1.4a) or (2.4): given any initial condition

(3.1.22) $$y^0 \in (H^{\frac{1}{2}+\epsilon}(\Omega))^d \cap H$$

show that there exists a boundary controller

(3.1.23) $$g \in L^2(0,\infty;(L^2(\Gamma))^d), \quad g \cdot \nu \equiv 0,$$

such that the corresponding solution of the (translated) linearized model (3.1.2), or (3.1.4a) satisfies the following regularity properties:

(3.1.24) $$y(\,\cdot\,;y^0) \in L^2\left(0,\infty;(H^{\frac{3}{2}+2\epsilon}(\Omega))^d \cap H\right), \text{ hence}$$

$$y_t(\,\cdot\,;y^0) = -P\mathbb{A}y \in L^2\left(0,\infty;(H^{-\frac{1}{2}+2\epsilon}(\Omega))^d \cap H\right);$$

hence [**L-M.1**, Theorem 3.1, p. 19]
(3.1.25)
$$y \in C\left([0,T];(H^{\frac{1}{2}+2\epsilon}(\Omega))^d \cap H\right), \text{ ultimately } y|_\Gamma = g \in C\left([0,T];(H^{2\epsilon}(\Gamma))^d\right).$$

[Problem (3.1.22)–(3.1.25) serves as the Finite Cost Condition for the Optimal Control Problem to be studied in Section 4, Eqns. (4.1.2)–(4.1.4).] Then, the trace regularity result in (3.1.25) imposes the boundary compatibility condition

(3.1.26) $$(y|_\Gamma)_{t=0} = y^0|_\Gamma = g(0).$$

Thus, if the boundary control g were finite-dimensional, we would have that only very special initial conditions y^0 could be covered, as to satisfy (3.1.26).

We have thus established the following result.

PROPOSITION 3.1.3. *Let $d = 3$. Solution of the local boundary stabilization problem of the N-S model (1.1) needs to be achieved in the topology of $(H^{\frac{3}{2}+2\epsilon}(\Omega))^3 \cap H$ in the space variable. This in turn requires a preliminary solution of the Finite Cost Condition (3.1.22)–(3.1.25). However, this FCC cannot be accomplished for all initial conditions as in (3.1.22), if the boundary control g is finite-dimensional.*

In other words: a finite-dimensional boundary control g cannot satisfy the FCC (3.1.22)–(3.1.25) (except for a meager set of very special Initial Conditions y^0); hence, it cannot produce local boundary stabilization of the original N-S model (1.1).

For this reason, for $d = 3$, we shall seek an infinite-dimensional boundary control g such that the FCC (3.1.22)–(3.1.25) is satisfied. We shall obtain such control g in Section 3.5.

Sharp regularity results for the linearized model (2.3), or its abstract version (3.1.4a). Henceforth, in this subsection we shall provide sharp (optimal) regularity results for the *non-homogeneous* linearized problem (2.3) [rewritten as (3.1.2)] or its abstract version (3.1.4a). The climax of these results is Theorem 3.1.8, the last one in this subsection. It is Theorem 3.1.8 for the special case $\theta = \frac{1}{2} + \frac{\epsilon}{2}$ that will be critically invoked in the proof of Theorem 3.5.1 below. The entire procedure may be divided in three parts: Part (a) provides Theorem 3.1.4 for the boundary datum g at the $H^{0,0}(\Sigma)$-level. Part (b) provides Theorem 3.1.7 for the boundary datum g at the $H^{2,1}(\Sigma)$-level. Finally, Part (c) interpolates between Parts (a) and (b) and yields the sought-after Theorem 3.1.8 for the boundary datum g (in particular, for $\theta = \frac{1}{2} + \frac{\epsilon}{2}$) at the $H^{1+\epsilon, \frac{1}{2}+\frac{\epsilon}{2}}(\Sigma)$-level, which is the one of interest in the proof of Theorem 3.5.1. Theorem 3.1.4 improves by (a critical) "ϵ" the semigroup proof of Lemma 3.1.1.

PART (A). THEOREM 3.1.4. *With reference to the abstract dynamics $y' - \mathcal{A}y = -\mathcal{A}Dg$ in (3.1.4a), we have with $L^2(0,T;(L^2(\Gamma))^d) \equiv H^{0,0}(\Sigma)$, $[\mathcal{D}(A^{\frac{1}{4}})]' = [\mathcal{D}((-\mathcal{A})^{\frac{1}{4}})]'$:*

(3.1.27a) $\quad \begin{cases} g \in H^{0,0}(\Sigma); \ y^0 \in [\mathcal{D}(A^{\frac{1}{4}})]' \Rightarrow y \in H^{\frac{1}{2},\frac{1}{4}}(Q) \cap H; \\ g \cdot \nu \equiv 0 \text{ on } \Sigma, \end{cases}$

(3.1.27b)

where we have defined

(3.1.28) $\quad H^{\frac{1}{2},\frac{1}{4}}(Q) \cap H \equiv L^2(0,T;(H^{\frac{1}{2}}(\Omega))^d \cap H) \cap H^{\frac{1}{4}}(0,T;H).$

PROOF. We follow, and modify accordingly, the proof given in [**L-M.1**, Vol. 2, p. 28], [**L-T.2**, Vol. 1, pp. 184–185] in the case of classical parabolic equations. The argument is based on three main steps.

STEP 1. LEMMA 3.1.5. *With reference to the abstract dynamics (3.1.4a), assume that*

(i)

(3.1.29) $\quad \begin{cases} g \in H^{\frac{3}{2},\frac{3}{4}}(\Sigma) \equiv L^2(0,T;(H^{\frac{3}{2}}(\Gamma))^d) \cap H^{\frac{3}{4}}(0,T;(L^2(\Gamma))^d); \\ g \cdot \nu \equiv 0 \text{ on } \Sigma, \end{cases}$

so that, a-fortiori, in view of [**L-M.1**, Theorem 3.1, p. 19, and p. 23, *for fractional derivatives, with $j = 0$, $m = \frac{3}{4}$, $(j+\frac{1}{2})/m = \frac{2}{3}$], we have*

(3.1.30) $\quad g \in C([0,T];(H^{\frac{1}{2}}(\Gamma))^d), \quad H^{\frac{1}{2}}(\Gamma) = [H^{\frac{3}{2}}(\Gamma), L^2(\Gamma)]_{\frac{2}{3}}.$

(ii)

(3.1.31) $\quad y^0 \in (H^1(\Omega))^d \cap H \text{ satisfying the C.C. } y^0|_\Gamma = g(0) \in (H^{\frac{1}{2}}(\Gamma))^d.$

Then, the corresponding solution of equation (3.1.4a) satisfies

(3.1.32) $\quad y \in H^{2,1}(Q) \cap H \equiv L^2(0,T;(H^2(\Omega))^d \cap H) \cap H^1(0;T;H).$

PROOF OF LEMMA 3.1.5. Thanks to the assumptions on the boundary datum g—both its regularity in (3.1.29) as well as the C.C. in (3.1.31)—there exists an interior extension g_e of the boundary term g, such that

$$(3.1.33) \qquad g_e \in H^{2,1}(Q) \cap H; \quad g_e|_\Gamma = g \in H^{\frac{3}{2},\frac{3}{4}}(\Sigma); \quad g_e|_{t=0} = y^0.$$

This is due to the surjectivity of the trace operator (in both time and space) under Compatibility Conditions: see [**L-M.1**, Vol. 2, page 28; Chapter 4, Section 2.5]. Now, however, in the case of the linearized N-S model (2.5)—and unlike the classical parabolic theory—we need to address the issue that, generally, $g_e \notin H$ since, generally, div $g_e \not\equiv 0$ in Ω. To this end, we then invoke [**Te.1**, Lemma 2.4, p. 32] or [**K.2**, Lemma 3.2.3, p. 134]: the divergence operator $v \to \text{div } v$ is an isomorphism of $\eta^\perp$ onto the space

$$(3.1.34a) \qquad L^2_0(\Omega) \equiv L^2(\Omega)/\mathcal{R} \equiv \left\{ f \in L^2(\Omega) : \int_\Omega f \, d\Omega = 0 \right\},$$

where $\eta^\perp$ is the orthogonal complement, in $(H^1_0(\Omega))^d$, for the scalar product

$$(3.1.34b) \qquad \int_\Omega \nabla u \cdot \nabla v \, d\Omega = \int_\Omega \sum_{i,j=1}^n \frac{\partial u_i}{\partial x_j} \frac{\partial v_i}{\partial x_j} \, d\Omega$$

of the space

$$(3.1.34c) \qquad W \equiv \{ f \in (H^1_0(\Omega))^d : \text{div } f = 0 \}.$$

To do this, we first observe that $f \equiv \text{div } g_e \in L^2_0(\Omega)$, since $\int_\Omega \text{div } g_e \, d\Omega = \int_\Gamma g \cdot \nu \, d\Gamma = 0$ by (3.1.29) ($g \cdot \nu \equiv 0$) and (3.1.33). Thus, we can invoke the aforementioned isomorphism result of the divergence operator at each instant t. Thus, for each $0 \leq t \leq T$, there is a unique vector $v \in W^\perp \subset (H^1_0(\Omega))^d$ such that

$$(3.1.35a) \qquad \text{div } v = \text{div } g_e \text{ in } \Omega; \quad v|_\Gamma = 0, \text{ for each } 0 \leq t \leq T;$$

in particular, for $t = 0$, recalling (3.1.33) and $y^0 \in H$ by (3.1.31)

$$(3.1.35b) \qquad \begin{cases} \text{div}(v|_{t=0}) = \text{div}(g_e|_{t=0}) = \text{div } y^0 = 0; \ (v|_{t=0})|_\Gamma = 0, \\ \text{and we can take } v|_{t=0} = 0. \end{cases}$$

By (3.1.33) on g_e, we have that

$$(3.1.36) \qquad \text{div } g_e \in H^{1,1}(Q), \text{ hence } v \in H^{2,1}(Q),$$

that is, the vector v in (3.1.35) preserves the regularity of g_e. We then set

$$(3.1.37) \qquad \psi \equiv g_e - v \in H^{2,1}(Q) \cap H \equiv L^2(0,T;(H^2(\Omega))^d \cap H) \cap H^1(0,T;H).$$

Recalling (3.1.35a–b), (3.1.33), we then have

$$(3.1.38) \qquad \text{div } \psi \equiv 0 \text{ in } \Omega; \ \psi|_\Gamma = g_e|_\Gamma = g; \ \psi|_{t=0} = g_e|_{t=0} - v|_{t=0} = y^0.$$

We next introduce a new variable $z \equiv y - \psi$ with y solution of problem (3.1.2), or (2.3). Then, z satisfies the following problem, by use also of (3.1.38)

$$(3.1.39a) \qquad \begin{cases} z_t + \mathbb{A}z = \nabla p + f & \text{in } \Omega \times (0,T]; \quad f \equiv -\mathbb{A}\psi - \psi_t; \\ (3.1.39b) \qquad \nabla \cdot z = 0 & \text{in } \Omega; \\ (3.1.39c) \qquad z(x,0) = 0 & \text{in } \Omega; \\ (3.1.39d) \qquad z = 0 & \text{in } \partial\Omega \times (0,T]. \end{cases}$$

We next project problem (3.1.39) on H by applying the Leray projector P and obtain the abstract version of (3.1.39), recalling $\mathcal{A}$ in (1.11):

(3.1.40) $$z_t = \mathcal{A}z + Pf, \qquad z(0) = 0,$$

where by (3.1.37) we have

(3.1.41) $$f \equiv -\mathbb{A}\psi - \psi_t \in L^2(0,T;(L^2(\Omega))^d); \quad Pf \in L^2(0,T;H).$$

Thus, standard parabolic or analytic semigroup theory yields

(3.1.42a) $$z(t) = \int_0^t e^{\mathcal{A}(t-\tau)} Pf(\tau) d\tau \in L^2(0,T;\mathcal{D}(\mathcal{A}));$$

(3.1.42b) $$z_t(t) = Pf(t) + \int_0^t \mathcal{A} e^{\mathcal{A}(t-\tau)} Pf(\tau) d\tau \in L^2(0,T;H),$$

or

(3.1.43) $$z \in H^{2,1}(Q) \cap H, \text{ so that } y = z + \psi \in H^{2,1}(Q) \cap H,$$

recalling again (3.1.37). Then (3.1.43) shows (3.1.32), as desired. □

STEP 2. LEMMA 3.1.6. *With reference to the abstract dynamics (3.1.4a), we have:*

(3.1.44) $$y^0 = 0, \ g \in H^{-\frac{1}{2},-\frac{1}{4}}(\Sigma) = [H^{\frac{1}{2},\frac{1}{4}}_{,0}(\Sigma)]' = [H^{\frac{1}{2},\frac{1}{4}}(\Sigma)]'$$
$$\Rightarrow Lg = y \in L^2(0,T;H) \equiv H^{0,0}(Q),$$

see [**L-M.1**, Vol. II, p. 41] *for the duality. [The subindex "0" means "vanishing at $t = 0$ and $t = T$" but it does not apply here, since the time regularity $\frac{1}{4}$, being below $\frac{1}{2}$, does not recognize these "boundary" conditions.]*

PROOF OF LEMMA 3.1.6. We show (3.1.44) by duality. That is, we equivalently show that

(3.1.45) $$L^* : L^2(0,T;H) \to H^{\frac{1}{2},\frac{1}{4}}(\Sigma).$$

In fact, let $f \in L^2(0,T;H)$. Recalling L in (3.1.6b) and $\mathcal{K}$ in (3.1.13b) (see also (3.1.72), (3.1.73) below), we compute

(3.1.46) $$(L^*f)(t) = -D^* \mathcal{A}^* \int_t^T e^{\mathcal{A}^*(\tau-t)} f(\tau) d\tau = -D^* \mathcal{A}^* (\mathcal{K}^* f)(t).$$

By proceeding as in proving (3.1.42a–b), we likewise obtain:

$$(\mathcal{K}^* f)(t) = \int_t^T e^{\mathcal{A}^*(\tau-t)} f(\tau) d\tau : \text{ continuous } L^2(0,T;H)$$

(3.1.47) $$\to L^2(0,T;\mathcal{D}(\mathcal{A})) \cap H^1(0,T;H) \subset H^{2,1}(Q).$$

Then, recalling the expression (3.2.2) for $D^* \mathcal{A}^*$ on $\mathcal{D}(\mathcal{A}^*) = \mathcal{D}(\mathcal{A})$, we rewrite (3.1.46), in view of (3.1.47) as

(3.1.48) $$(L^* f)(t) = -\nu_0 \frac{\partial}{\partial \nu} (\mathcal{K}^* f)(t).$$

We now apply trace theory (in time and space) as in [**L-M.1**, Vol. 2, p. 9] to (3.1.42), (3.1.48), to obtain

(3.1.49) $$L^* f = -\nu_0 \frac{\partial}{\partial \nu} \mathcal{K}^* f \in H^{\frac{1}{2},\frac{1}{4}}(\Sigma),$$

and (3.1.45) is established. Lemma 3.1.6 is proved. □

Step 3. To complete the proof of Theorem 3.1.4, we now interpolate [**L-M.1**, Vol. 1, p. 27] between the regularity $g \to y$ in Lemma 3.1.5 (see (3.1.29), (3.1.32)) and the regularity $g \to y$ in (3.1.44) of Lemma 3.1.6, where the C.C. is irrelevant at $\theta = \frac{3}{4}$; that is, for $g \in L^2(\Sigma)$:

$$(3.1.50) \quad g \in L^2(\Sigma) = \left[H^{\frac{3}{2},\frac{3}{4}}(\Sigma), H^{-\frac{1}{2},-\frac{1}{4}}(\Sigma)\right]_{\theta=\frac{3}{4}}$$

$$\Rightarrow Lg \in [H^{2,1}(Q), H^{0,0}(Q)]_{\theta=\frac{3}{4}} = H^{\frac{1}{2},\frac{1}{4}}(Q),$$

as desired. The proof of the regularity result (3.1.27) of Theorem 3.1.4 is complete.

PART (B). THEOREM 3.1.7. *With reference to the abstract dynamics* $y' - \mathcal{A}y = -\mathcal{A}Dg$ *in (3.1.4a), assume that*
(i)

$$(3.1.51) \quad \begin{cases} g \in H^{2,1}(\Sigma) \equiv L^2(0,T; (H^2(\Gamma))^d) \cap H^1(0,T; (L^2(\Gamma))^d); \\ g \cdot \nu \equiv 0 \text{ on } \Sigma, \end{cases}$$

so that, a fortiori, in view of [**L-M.1**, *Theorem 3.1, p. 19, with* $j = 0$, $m = 1$], *we have*

$$(3.1.52) \quad g \in C([0,T]; (H^1(\Gamma))^d)), \quad H^1(\Gamma) = [H^2(\Gamma), H^0(\Gamma)]_{\frac{1}{2}};$$

(ii)

$$(3.1.53) \quad y^0 \in (H^{\frac{3}{2}}(\Omega))^d \cap H \text{ satisfying the C.C. } y^0|_\Gamma = g(0) \in (H^1(\Gamma))^d.$$

Then, the corresponding solution of Eqn. (3.1.4a) satisfies

$$(3.1.54) \quad y \in H^{\frac{5}{2},\frac{5}{4}}(Q) \equiv L^2(0,T; (H^{\frac{5}{2}}(\Omega))^d \cap H) \cap H^{\frac{5}{4}}(0,T; H).$$

PROOF OF THEOREM 3.1.7. We use the solution formula (3.1.13)

$$(3.1.55) \quad y(t) = e^{\mathcal{A}t}[y^0 - Dg(0)] + Dg(t) - \int_0^t e^{\mathcal{A}(t-\tau)} Dg_t(\tau) d\tau.$$

Then, with $g_t \in L^2(0,T; (L^2(\Gamma))^d)$, $g_t \cdot \nu \equiv 0$ on Σ, by assumption (3.1.51), we invoke (3.1.27) of Theorem 3.1.4 and obtain (recall L in (3.1.6b)):

$$(3.1.56a) \quad q \equiv \mathcal{A} \int_0^t e^{\mathcal{A}(t-\tau)} Dg_t(\tau) d\tau \in H^{\frac{1}{2},\frac{1}{4}}(Q) \cap H,$$

hence *a-fortiori*,

$$(3.1.56b) \quad y_2(t) \equiv \int_0^t e^{\mathcal{A}(t-\tau)} Dg_t(\tau) d\tau = \mathcal{A}^{-1} q \in L^2(0,T; \mathcal{D}(\mathcal{A})),$$

or $\mathcal{A}y_2 = q$. Hence, recalling $\mathbb{A}$ from (3.1.1), we can rewrite this as (at each t, a.e.):

$$(3.1.57a) \qquad \begin{cases} \mathbb{A}y_2 = q \in H^{\frac{1}{2},\frac{1}{4}}(Q); \\ (3.1.57b) \qquad \text{div } y_2 \equiv 0 \text{ in } \Omega; \\ (3.1.57c) \qquad y_2|_\Gamma \equiv 0 \text{ in } \Gamma. \end{cases}$$

The known regularity result [**C-F.1**], [**Te.1**, p. 33] of the Stokes problem yields then that the solution y_2 of (3.1.57) satisfies

(3.1.58) $$y_2(t) = \int_0^t e^{\mathcal{A}(t-\tau)} Dg_t(\tau)d\tau \in L^2(0,T;(H^{\frac{5}{2}}(\Omega))^d \cap H),$$

with a gain of two units in space regularity over q. Moreover, again recalling the space regularity (3.1.51) for g, we obtain via (3.1.3d)

(3.1.59) $$Dg \in L^2(0,T;(H^{\frac{5}{2}}(\Omega))^d \cap H).$$

Finally, assumption (3.1.53) yields y^0, $Dg(0) \in (H^{\frac{3}{2}}(\Omega))^d \cap H$, the latter via the regularity (3.1.3d) of D applied to $g(0) \in (H^1(\Gamma))^d$ by (3.1.52); ultimately, by the c.c. in (3.1.53),

(3.1.60) $$[y^0 - Dg(0)] \in (H^{\frac{3}{2}}(\Omega))^d \cap V \equiv \mathcal{D}(A^{\frac{3}{4}}) \equiv \mathcal{D}((-\mathcal{A})^{\frac{3}{4}}),$$

recalling (1.15). Then, (3.1.60) implies [recall, e.g., [**B-T.1**, Appendix A]]

$$y_1(t) \equiv e^{\mathcal{A}t}[y^0 - Dg(0)] = (-\mathcal{A})^{-\frac{3}{4}} e^{\mathcal{A}t}(-\mathcal{A})^{\frac{3}{4}}[y^0 - Dg(0)]$$

(3.1.61) $$\in L^2(0,T;\mathcal{D}((-\mathcal{A})^{\frac{5}{4}})) \subset L^2(0,T;(H^{\frac{5}{2}}(\Omega))^d \cap H),$$

since $\frac{3}{4} + \frac{1}{2} = \frac{5}{4}$. Using (3.1.58), (3.1.59), (3.1.61) in (3.1.13) yields

(3.1.62) $$y(t) = y_1(t) + Dg(t) - y_2(t) \in L^2(0,T;(H^{\frac{5}{2}}(\Omega))^d \cap H),$$

as desired. Thus, half of (3.1.54) (regularity in space) is now proved. To show the other half of (3.1.54) (regularity in time), we differentiate (3.1.55) in time and obtain, after a cancellation of $Dg_t(t)$,

(3.1.63) $$y_t(t) = \mathcal{A}y_1(t) - q(t) = \mathcal{A}e^{\mathcal{A}t}[y^0 - Dg(0)] - \mathcal{A}\int_0^t e^{\mathcal{A}(t-\tau)}Dg_t(\tau)d\tau.$$

From (3.1.61), we obtain

(3.1.64a) $\mathcal{A}y_1 \in L^2(0,T;\mathcal{D}((-\mathcal{A})^{\frac{1}{4}}));\ D_t(\mathcal{A}y_1) = \mathcal{A}^2 y_1 \in L^2(0,T;[\mathcal{D}((-\mathcal{A})^{\frac{3}{4}})]').$

By the intermediate derivative theorem [**L-M.1**, Vol. 1, p. 15, $m = 1$], applied to (3.1.64a), we obtain

(3.1.64b) $$D_t^{\frac{1}{4}}(\mathcal{A}y_1) \in L^2(0,T;H) \text{ since } \frac{1}{4}(1-\theta) - \frac{3}{4}\theta = 0, \text{ for } \theta = \frac{1}{4}.$$

Thus, (3.1.64b) on $\mathcal{A}y_1$ and (3.1.56a) on q yield via (3.1.63):

(3.1.65) $$\mathcal{A}y_1, q,\ y_t = \mathcal{A}y_1 - q \in H^{\frac{1}{4}}(0,T;H), \text{ or } y \in H^{\frac{5}{4}}(0,T;H),$$

and (3.1.65)—along with (3.1.62)—completes the proof of (3.1.54). Theorem 3.1.7 is established. $\square$

PART (c). By interpolation between Theorem 3.1.4 and Theorem 3.1.7, we obtain the following result, a special case of which—for $\theta = \frac{1}{2} + \frac{\epsilon}{2}$—will be critically invoked in the proof of Theorem 3.5.1 below.

THEOREM 3.1.8. *With reference to the abstract dynamics $y' - \mathcal{A}y = -\mathcal{A}Dg$ in (3.1.4a), assume that*
(i)

(3.1.66) $$g \in H^{2\theta,\theta}(\Sigma),\ 0 < \theta \le 1,\ g \cdot \nu \equiv 0 \text{ on } \Sigma,$$

so that, a-fortiori, *in view of* [**L-M.1**, Theorem 3.1, p. 19 in fractional form with $j = 0$, $m = \theta$, $(j + \frac{1}{2})/m = \frac{1}{2\theta}$, $\theta > 0$]

(3.1.67) $\quad g \in C([0,T]; (H^{2\theta-1}(\Omega))^d \cap H), \quad \frac{1}{2} \leq \theta \leq 1; \ H^{2\theta-1} = [H^{2\theta}, H^0]_{\frac{1}{2\theta}};$

(ii)

(3.1.68) $\quad\quad\quad\quad y^0 \in (H^{2\theta-\frac{1}{2}}(\Omega))^d \cap H, \ \frac{1}{4} \leq \theta \leq 1,$

$\quad\quad\quad\quad y^0 \in (H^{2\theta-\frac{1}{2}}(\Omega))^d = [(H^{\frac{1}{2}-2\theta}(\Omega))^d \cap H]', 0 < \theta < \frac{1}{4},$

satisfying the c.c.

(3.1.69) $\quad\quad\quad\quad y^0|_\Gamma = g(0) \in (H^{2\theta-1}(\Gamma))^d, \ \frac{1}{2} < \theta < 1.$

Then, the corresponding solution of Eqn. (3.1.4a) satisfies

(3.1.70) $\quad y \in H^{\frac{1}{2}+2\theta, \frac{1}{4}+\theta}(Q) \equiv L^2(0, T; (H^{\frac{1}{2}+2\theta}(\Omega))^d \cap H) \cap H^{\frac{1}{4}+\theta}(0, T; H).$

PROOF. We interpolate [**L-M.1**, Vol. 1, p. 27] between Theorem 3.1.4 for the subspace of the initial conditions $y^0 \in (H^{-\frac{1}{2}}(\Omega))^d \equiv [(H^{\frac{1}{2}}(\Omega))^d \cap H]'$, as $\mathcal{D}(A^{\frac{1}{4}}) \subset (H^{\frac{1}{2}}(\Omega))^d \cap H$, and Theorem 3.1.7, while keeping the c.c. (which are not recognized in Theorem 3.1.5), thereby obtaining Theorem 3.1.8. $\square$

REMARK 3.1.1. The sharp (optimal) regularity results of the mixed (Initial/Boundary Value) problem for the linearized equation (3.1.1) or (3.1.4a), given in Theorem 3.1.4 through Theorem 3.1.8, are of interest in their own right. They are expected to be useful in a variety of other situations. Even in the case of *classical* (linear) parabolic mixed problems, we can find optimal regularity results in some ranges of the interpolation parameter (such as to cover Theorem 3.1.4, for instance), but not in all ranges needed here. In particular, we cannot find the critical (for us, final) result in Theorem 3.1.8 (see regularity in (3.5.59) requiring $2\theta = 1 + \epsilon$) in comprehensive treaties such as [**L-M.1**, Vol. 2, "Final results ...," p. 78, as well as results at p. 65]. Our case in Theorem 3.1.8 falls in the range of "exceptional parameters" mentioned in [**L-M.1**, Vol. 2, p. 79, middle] where technical difficulties are out of proportion according to [**L-M.1**, Vol. 2, p. 87].

For completeness, we close this section, by complementing the regularity of L^* in (3.1.45) over a larger scale, thus obtaining the counterpart of [**L-T.2**, Lemma 3.2.3, p. 188] in the classical parabolic case.

THEOREM 3.1.9. *With reference to the operator L^* in (3.1.46), we have:*

(3.1.71) $\quad\quad L^* : \text{continuous } H^{2\theta, \theta}(Q) \cap H \rightarrow H^{2\theta+\frac{1}{2}, \theta+\frac{1}{4}}(\Sigma).$

PROOF. We interpolate between (3.1.45) for $\theta = 0$ and the statement for $\theta = 1$:

(3.1.72) $\quad\quad L^* : \text{continuous } H^{2,1}(Q) \cap H \rightarrow H^{\frac{5}{2}, \frac{5}{4}}(\Sigma),$

thereby obtaining (3.1.71). The proof of (3.1.72) proceeds as in the classical case [**L-T.2**, p. 192], by integration by parts, using again (3.2.2): $D^*A^* = \nu_0 \frac{\partial}{\partial \nu}$ and the standard trace theory (in time and space) of [**L-M.1**, Vol. 2, p. 9]. The improvement by "ϵ" over the classical parabolic general case [**L-T.2**, Lemma 3.2.3, Eqn. (3.2.13), p. 188] is because now, for the present linearized operator $\mathcal{A} = (-\nu_0 A + A_0)$ in

(1.11)—which is a perturbation of a self-adjoint operator by a lower-order A_0, $\mathcal{D}(A_0) = \mathcal{D}(A^{\frac{1}{2}}))$, we have the validity of regularity

(3.1.73) $\qquad (-\mathcal{A})^{\frac{1}{2}} e^{\mathcal{A}t} :$ continuous $H \to L^2(0, T; H)$,

and similar ones [**B-T.1**, Appendix], already tacitly used in (3.1.8). □

The operators L_t, $\mathcal{K}$ and their adjoints L_t^* and $\mathcal{K}^*$. We close this subsection by collecting operators arising from (3.1.6b) and (3.1.13b), which will be invoked in the Riccati theory treatment of Section 4. In Section 4, we can assume w.l.o.g. that the s.c. semigroup $e^{\mathcal{A}t}$ is exponentially stable on H. Thus, under this *assumption*, we shall refine the operator in (3.1.6b) and introduce, for $g \in L^2(0, \infty; (L^2(\Gamma))^d)$, $g \cdot \nu \equiv 0$ on Σ:

$$(L_s g)(t) \equiv -\mathcal{A} \int_s^t e^{\mathcal{A}(t-\tau)} Dg(\tau) d\tau, \ 0 \leq s \leq t$$

(3.1.74) $\qquad\qquad :$ continuous $L^2(0, \infty; (L^2(\Gamma)^d) \to H^{\frac{1}{2}, \frac{1}{4}}(Q_\infty) \cap H$,

whose $L^2(0, \infty; H)$-adjoint is

$$(L_s^* f)(t) \equiv -D^* \mathcal{A}^* \int_t^\infty e^{\mathcal{A}^*(\tau-t)} f(\tau) d\tau, \ 0 \leq s \leq t$$

(3.1.75) $\qquad\qquad :$ continuous $H^{2\theta, \theta}(Q_\infty) \to H^{2\theta + \frac{1}{2}, \theta + \frac{1}{2}}(\Sigma_\infty)$,

by Theorem 3.1.8 and Theorem 3.1.9. Moreover, the $L^2(0, \infty; H)$-adjoint of the operator $\mathcal{K}$ in (3.1.13b) is

(3.1.76) $\qquad (\mathcal{K}^* h)(t) = \int_t^\infty e^{\mathcal{A}^*(\tau-t)} h(\tau) d\tau \in L^2(0, \infty; \mathcal{D}(\mathcal{A}^*))$.

3.2. The operator $D^* \mathcal{A}^*$, $D^* : H \to (L^2(\Gamma))^d$

It is not difficult to show the well-known result that the H-dual $\mathcal{A}^* = -(\nu_0 A + A_0^*)$ of the operator $\mathcal{A} = -(\nu_0 A + A_0)$ in (1.11) is explicitly given by

(3.2.0)

$$\mathcal{A}^* f = -(\nu_0 A + A_0^*) f$$

$$= -\nu_0 P \left[\Delta f + (y_e \cdot \nabla) f - \begin{bmatrix} \sum_{j=1}^d (\partial_1 y_{ej}) f_j \\ \cdots \\ \sum_{j=1}^d (\partial_d y_{ej}) f_j \end{bmatrix} \right], \ f \in \mathcal{D}(\mathcal{A}^*) = \mathcal{D}(\mathcal{A}).$$

One uses also the notation $(\nabla y_e)^T f$. We shall not use this explicit expression, however. The goal of the present subsection is to establish the following result on $D^* \mathcal{A}^*$ (in line with [**L-T.1**], [**L-T.2**] in the classical parabolic case such as the heat equation. To this end, we shall make critical use of the *pointwise* c.c. $g \cdot \nu \equiv 0$ on Σ (3.1.2e), but we shall not use (3.2.0) explicitly. Here $D^* : H \to (L^2(\Gamma))^d$ is defined by

(3.2.1) $\quad (Df, h)_H = (f, D^* h)_{(L_2(\Gamma))^d}, \quad f \in (L_2(\Gamma))^d, \quad f \cdot \nu \equiv 0 \text{ on } \Gamma, \ h \in H,$

so that $Df \in (H^{\frac{1}{2}}(\Omega))^d \cap H$, by (3.1.3d). Thus, we can apply D^* on $\mathcal{A}^* v \in H$, $v \in \mathcal{D}(\mathcal{A}^*)$.

PROPOSITION 3.2.1. *Let $v \in \mathcal{D}(\mathcal{A}^*) = \mathcal{D}(\mathcal{A}) = (H^2(\Omega))^d \cap V \subset H$, see (1.11). Then, with reference to (3.1.3d), (3.2.0), (3.2.1), we have*

$$(3.2.2) \qquad D^* \mathcal{A}^* v = \nu_0 \frac{\partial v}{\partial \nu}.$$

[Lemma 3.3.1 below will then show that $\frac{\partial v}{\partial \nu} \cdot \nu = 0$ on Γ, for $v \in \mathcal{D}(\mathcal{A}^*)$.]

PROOF. This is based on the following steps, where we avoid use of (3.2.0).

Step 1. Let $v \in \mathcal{D}(\mathcal{A}^*)$, $g \in (L_2(\Gamma))^d$, $g \cdot \nu \equiv 0$ on Γ, so that $Dg \in (H^{\frac{1}{2}}(\Omega))^d \cap H$ by (3.1.3d). Then, by (3.2.1) and $-\mathcal{A}^* = \nu_0 A + A_0^*$, we have

$$(3.2.3) \quad -(D^* \mathcal{A}^* v, g)_{(L_2(\Gamma))^d} = ((-\mathcal{A}^*)v, Dg)_H = \nu_0 (Av, Dg)_H + (A_0^* v, Dg)_H.$$

Step 2. Let v, g be as in Step 1. Here we shall show that

$$(3.2.4) \qquad (Av, Dg)_H = -\int_\Omega v \Delta(Dg) d\Omega - \int_\Gamma \frac{\partial v}{\partial \nu} g \, d\Gamma.$$

In fact, recalling A in (1.6), the orthogonal Leray projector $P = P^*$ onto H, and $Dg \in H$ as $g \cdot \nu = 0$ on Γ, so that $PDg = Dg$, we obtain via Green's second theorem

$$(3.2.5) \quad (Av, Dg)_H = -(P\Delta v, Dg)_H = -(\Delta v, PDg)_H = -(\Delta v, Dg)_H$$

$$= -\int_\Omega \Delta v (Dg) d\Omega = -\int_\Omega v \Delta(Dg) d\Omega$$

$$- \int_\Gamma \frac{\partial v}{\partial \nu} g \, d\Gamma + \int_\Gamma v \frac{\partial(Dg)}{\partial \nu} d\Gamma,$$

where in (3.2.5) we have used $Dg|_\Gamma = g$ by (3.1.3c), and $v|_\Gamma = 0$ for $v \in \mathcal{D}(\mathcal{A}^*)$. Then, (3.2.5) yields (3.2.4) as desired.

REMARK 3.2.1. If we relaxed the boundary c.c. on g to the integral version $\int_\Gamma g \cdot \nu \, d\Gamma \equiv 0$ in (3.1.2e)—so that $Dg \notin H$, but $PDg \in H$—and attempted to compute $(Av, PDg)_H$, we would run into a problem: the counterpart of (3.2.5) would then be

$$(Av, PDg)_H = -\int_\Omega v \Delta(PDg) d\Omega - \int_\Gamma \frac{\partial v}{\partial \nu} ([PDg]_\Gamma) d\Gamma,$$

where now $[PDg]_\Gamma \neq g$, since $[PDg]_\Gamma \cdot \nu = 0$, while generally $g \cdot \nu \neq 0$ by selection. Thus, (3.2.2) *cannot* be extended to read: $D^* P \mathcal{A}^* v = \frac{\partial v}{\partial \nu}$, if D only satisfies (3.1.3e) and not (3.1.3d).

Step 3. Recall A_0 from (1.10) and write for $h \in H$, $\nabla h \in (H^{-1}(\Omega))^d$ and $v \in V = (H_0^1(\Omega))^d \cap H$, see (1.3), so that all terms below are well defined as duality pairings, while $Pv = v$ for the orthogonal Leray projector $P = P^*$:

$$
\begin{aligned}
(3.2.6) \quad (A_0 h, v)_H &= (P[(y_e \cdot \nabla) h], v)_H + (P[(h \cdot \nabla) y_e], v)_H \\
&= ((y_e \cdot \nabla) h, v)_H + ((h \cdot \nabla) y_e, v)_H \\
&= (h, A_0^* v)_H, \quad h \in H, \quad v \in V.
\end{aligned}
$$

Specializing (3.2.6) with $h = Dg \in (H^{\frac{1}{2}}(\Omega))^d \cap H$, see (3.1.3d), for $g \in (L_2(\Gamma))^d$, $g \cdot \nu \equiv 0$ on Γ, we obtain still with $v \in V$:

$$
(3.2.7) \quad (A_0^* v, Dg)_H = ((y_e \cdot \nabla) Dg + (Dg \cdot \nabla) y_e, v)_H.
$$

Step 4. We now substitute (3.2.4) and (3.2.7) in (3.2.3), thus obtaining for $v \in \mathcal{D}(\mathcal{A}^*) = \mathcal{D}(\mathcal{A}) = \mathcal{D}(A)$ and $g \in (L_2(\Gamma))^d$, $g \cdot \nu \equiv 0$ on Γ:

$$
\begin{aligned}
(3.2.8) \quad &-(D^* \mathcal{A}^* v, g)_{(L_2(\Gamma))^d} \\
&= (v, -\nu_0 \Delta(Dg) + (y_e \cdot \nabla) Dg + (Dg \cdot \nabla) y_e)_H - \nu_0 \int_\Gamma \frac{\partial v}{\partial \nu} g \, d\Gamma \\
\text{(by (3.1.1))} \quad &= (v, \mathbb{A}(Dg))_H - \nu_0 \left(\frac{\partial v}{\partial \nu}, g \right)_{(L_2(\Gamma))^d},
\end{aligned}
$$

where $\mathbb{A}\psi = 0$, $\psi = Dg$ in (3.2.8), by (3.1.3a). Then (3.2.8) proves (3.2.2), as desired. □

3.3. A critical boundary property related to the boundary c.c. in (3.1.2e)

We now provide a critical property, which will have the implication to assert that for the sought-after boundary control g needed to stabilize the linearized problem (3.1.2), we do have the required property $g \cdot \nu \equiv 0$ on Σ in (3.1.2e): see Section 3.5, Eqn. (3.5.47) in the general case, as well as Lemma 3.6.1 for $g = \sum_{k=1}^{K} u_k(t) w_k$, under the FDSA = (3.6.2), where $w_i \in \text{span}\{\partial_\nu \varphi_i^*\}_{j=1}^{N}$, φ_j^* being an eigenfunction of $\mathcal{A}^*$.

LEMMA 3.3.1. *Let $\varphi \in (C^1(\Omega))^d$ be a function satisfying the following two properties:*
(3.3.1)
(i) $\varphi|_\Gamma \equiv 0$; (ii) $\nabla \cdot \varphi \equiv 0$ *in $\bar{\Omega}$ (actually only on an interior strip of Γ).*

Then we have:
(3.3.2a)
$$(\nabla \varphi \cdot \nu) \cdot \nu \equiv 0 \text{ on } \Gamma; \text{ thus, the boundary vector } \frac{\partial \varphi}{\partial \nu} = \nabla \varphi \cdot \nu \text{ is tangential to } \Gamma, \text{ i.e., is}$$
orthogonal to the normal ν on Γ.

In particular, via (3.2.2), the vector

$$
(3.3.2b) \quad D\mathcal{A}^* v = \nu_0 \frac{\partial v}{\partial \nu} \text{ is tangential on } \Gamma, \quad v \in \mathcal{D}(\mathcal{A}^*) \subset V.
$$

PROOF. For $\varphi = [\varphi_1, \ldots, \varphi_d]$, condition (i) implies that each gradient $\nabla \varphi_i$ is parallel to ν, $i = 1, \ldots, d$. Thus, we may write on Γ:

(3.3.3) $\quad$ on $\Gamma: \nabla \varphi_i = s_i |\nabla \varphi_i| \nu; \quad \partial_i \varphi_i = s_i |\nabla \varphi_i| \nu_i, \quad i = 1, \ldots, d, \quad \partial_i = \partial_{x_i}$,

where the symbol s_i stands for either $+1$ or else -1 (according to as whether the level surfaces $\varphi_i \equiv \epsilon > 0$ are external or internal to Γ: here ν, as usual, is the unit outward normal vector to Γ). Moreover, condition (ii)—when restricted to Γ—yields, by virtue of (3.3.3):

$$(3.3.4) \qquad \text{div } \varphi = \nabla \cdot \varphi = \sum_{i=1}^{d} \partial_i \varphi_i = \sum_{i=1}^{d} s_i |\nabla \varphi_i| \nu_i \equiv 0 \text{ on } \Gamma.$$

Next, we compute on Γ, by invoking (3.3.3) (left),

$$(3.3.5) \qquad (\nabla \varphi \cdot \nu) \cdot \nu \stackrel{\text{def}}{=} \begin{bmatrix} \nabla \varphi_1 \cdot \nu \\ \vdots \\ \nabla \varphi_d \cdot \nu \end{bmatrix} \cdot \nu \stackrel{\text{def}}{=} \sum_{i=1}^{d} \nu_i (\nabla \varphi_i \cdot \nu)$$

$$= \sum_{i=1}^{d} \nu_i s_i |\nabla \varphi_i| \nu \cdot \nu = \sum_{i=1}^{d} \nu_i s_i |\nabla \varphi_i| \equiv 0,$$

where in the last steps we have used $\nu \cdot \nu = 1$ and (3.3.4). Thus, (3.3.5) establishes (3.3.2). $\square$

Lemma 3.3.1 will be applied to our stabilization problem in Section 3.5 (Eqn. (3.5.47) and in Section 3.6, Lemma 3.6.1.

3.4. Some technical preliminaries; space and system decomposition

Complexification and projections. We shall henceforth denote by H and V the complexified spaces $H \oplus iH$, $V \oplus iV$ and consider the extension of the linearized system (3.1.4a) of the translated problem (3.1.2) or (2.3) to these spaces, thus obtaining

$$(3.4.1) \qquad \frac{dz}{dt} - \mathcal{A}z = -\mathcal{A}Dv \in [\mathcal{D}(\mathcal{A}^*)]', \ t \geq 0, \ z(0) = z_0, \ v \cdot \nu = 0 \text{ on } \Sigma.$$

where $z = y + i\tilde{y}$, $v = g + i\tilde{g}$, $z_0 = y^0 + i\tilde{y}^0$, and where the boundary datum v satisfies the pointwise c.c. $v \cdot \nu \equiv 0$ in Σ, so that $Dv \in H$, as we have seen in (3.1.3d). The operator $\mathcal{A}$ in (3.4.1) is actually the *extension*, by transposition, $\mathcal{A}: H \to [\mathcal{D}(\mathcal{A}^*)]'$ of the original operator $\mathcal{A}$ in (1.11). This was noted below (1.12), (2.2), and below (2.4). Moreover, as noted below (1.11), the operator $\mathcal{A}$ has a compact resolvent in H, and generates an analytic C_0-semigroup on H, so that only a finite number of eigenvalues of $\mathcal{A}$ are in the right complex half-plane $\{\lambda \in \mathbb{C}; \text{Re } \lambda \geq 0\}$ see (1.12). We have already denoted by $\lambda_1, \lambda_2, \ldots, \lambda_N$ these (unstable) eigenvalues of $\mathcal{A}$ repeated according to their algebraic multiplicity ℓ_i so that

$$(3.4.2) \qquad \text{Re } \lambda_{N+1} < 0 \leq \text{Re } \lambda_N \leq \cdots \leq \text{Re } \lambda_1.$$

We have also let M be the number of distinct unstable eigenvalues λ_j of $\mathcal{A}$.

3.4. SOME TECHNICAL PRELIMINARIES

Space decomposition. Denote by P_N and P_N^* the projections

$$(3.4.3) \quad P_N = -\frac{1}{2\pi i}\int_\Gamma (\lambda I - \mathcal{A})^{-1}d\lambda; \qquad P_N^* = -\frac{1}{2\pi i}\int_{\overline{\Gamma}}(\lambda I - \mathcal{A}^*)^{-1}d\lambda$$

$$: H \stackrel{\text{onto}}{\to} Z_N^u; \qquad\qquad\qquad\qquad : H \stackrel{\text{onto}}{\to} (Z_N^u)^*$$

explicitly identified in [**K.1**, p. 178] as a contour integral, where, in the case of $P_N : H$ onto Z_N^u, Γ separates the unstable spectrum from the stable one of $\mathcal{A}$, and similarly for P_N^*: H onto $(Z_N^u)^*$ where $\overline{\Gamma}$ is the complex conjugate of Γ. Then, the (complexified) space H can be decomposed in complementary, not necessarily orthogonal, subspaces, as (see [**K.1**, p. 178])

$$(3.4.4) \quad H = Z_N^u \oplus Z_N^s; \quad Z_N^u \equiv P_N H; \quad Z_N^s = (I - P_N)H; \quad \dim Z_N^u = N,$$

where each of the spaces Z_N^u and Z_N^s is invariant under $\mathcal{A}$. We set

$$(3.4.5) \quad \mathcal{A}_N^u = P_N \mathcal{A} = \mathcal{A}|_{Z_N^u}; \quad \mathcal{A}_N^s = (I - P_N)\mathcal{A} = \mathcal{A}|_{Z_N^s}$$

for the restrictions of $\mathcal{A}$ to Z_N^u and Z_N^s, respectively. We then have that the spectra of $\mathcal{A}$ on Z_N^u and Z_N^s coincide with $\{\lambda_j\}_{j=1}^N$ and $\{\lambda_j\}_{j=N+1}^\infty$, respectively:

$$(3.4.6) \quad \sigma(\mathcal{A}_N^u) = \{\lambda_j\}_{j=1}^N; \qquad \sigma(\mathcal{A}_N^s) = \{\lambda_j\}_{j=N+1}^\infty.$$

Moreover, since $\mathcal{A}$ generates a C_0-analytic semigroup on H, then its restriction $\mathcal{A}_N^s$ to Z_N^s generates likewise a C_0-analytic semigroup on Z_N^s. This implies that $\mathcal{A}_N^s$ satisfies the spectrum determined growth condition on Z_N^s (see [**T.1**]), and so we have

$$(3.4.7) \quad \left\|e^{\mathcal{A}_N^s t}\right\|_{\mathcal{L}(H;H)} \leq C_{\gamma_0}e^{-\gamma_0 t}, \quad \forall\, t \geq 0,$$

$$\left\|(-\mathcal{A}_N^s)^\theta e^{\mathcal{A}_N^s t}\right\|_{\mathcal{L}(H)} \leq C\frac{e^{-\gamma_0 t}}{t^\theta}, \quad 0 < \theta < 1,\ t > 0,$$

for any $0 < \gamma_0 < |\text{Re }\lambda_{N+1}|$, where C_{γ_0} depends on γ_0.

System decomposition. Then system (3.4.1) with $v \cdot \nu = 0$ on Σ can accordingly be decomposed as

$$(3.4.8) \quad z = z_N + \zeta_N, \quad z_N = P_N z, \quad \zeta_N = (I - P_N)z_N,$$

where applying P_N and $(I - P_N)$ (which commute with $\mathcal{A}$) on (3.4.1), we obtain via (3.4.5)

$$(3.4.9) \quad \text{on } Z_N^u: \quad z_N' - \mathcal{A}_N^u z_N = -P_N(\mathcal{A}Dv) = -\mathcal{A}_N^u P_N Dv,$$

$$z_N(0) = P_N z_0,$$

$$(3.4.10) \quad \text{on } Z_N^s: \quad \zeta_N' - \mathcal{A}_N^s \zeta_N = -(I - P_N)(\mathcal{A}Dv) = -\mathcal{A}_N^s(I - P_N)Dv,$$

$$\zeta_N(0) = (I - P_N)z_0,$$

respectively. In (3.4.9), (3.4.10), actually P_N is the extension of (3.4.3) from H to $[\mathcal{D}(\mathcal{A}^*)]'$. Some explicit representations will be given below in (3.6.7) and (3.6.8). In Section 3.6 below, we study the feedback stabilization of the finite-dimensional unstable system (3.4.9), with spectrum given by (3.4.6) (left).

3.5. Theorem 2.1, general case $d = 3$: An infinite-dimensional open-loop boundary controller g satisfying the FCC (3.1.22)–(3.1.24) for the linearized system (3.1.4): $g \in L^2(0,\infty;(L^2(\Gamma))^3)$, $g \cdot \nu \equiv 0$ on Σ; $y^0 \in (H^{\frac{1}{2}+\epsilon}(\Omega))^3 \cap H \Rightarrow y \in L^2(0,\infty;(H^{\frac{3}{2}+\epsilon}(\Omega))^3 \cap H)$

Orientation. We have already seen in Proposition 3.1.3 that, for $d = 3$, the Finite Cost Condition (3.1.22)–(3.1.25) for the linearized system (3.1.4) cannot be satisfied for all I.C. (initial conditions) within the class of open-loop finite-dimensional controls. This still leaves open the question as to whether such FCC for (3.1.4) can be satisfied at all for all I.C.—therefore by necessity within the class of open-loop infinite-dimensional controls. We shall establish in this section that the answer is in the affirmative. The proof below will be based on the exponential stabilization result of the *linearized system* by finite-dimensional *interior* controllers (with arbitrarily small interior support) recently obtained in [**B-T.1**]. The result that we shall obtain is fully general. However, for general I.C., it requires that the sought-after boundary control g acts on the entire boundary $\Gamma = \partial\Omega$. See also Remark 3.5.1.

In subsequent Sections 3.6 and 3.7, we shall see that a radically different approach will allow for the required open-loop boundary control g to possess additional desirable properties including the following two: (i) to be active only on an arbitrarily small portion Γ_1 (of positive surface measure) of the boundary $\Gamma = \partial\Omega$; (ii) to enjoy a finite-dimensional character, in the sense that the difference $[g - e^{-2\gamma_1 t} y^0|_{\Gamma}]$ be finite-dimensional. However, these additional features come at a price, in that the alternative approach which produces them requires a "finite-dimensional spectral assumption," see FDSA = (3.6.2) below (which we believe to be "generically" true with respect to the class of bounded domains Ω).

THEOREM 3.5.1. *Let $d = 3$. Let Ω be connected. There exists a infinite-dimensional open-loop boundary controller g*

$$(3.5.1) \qquad g \in L^2(0,\infty;(L^2(\Gamma))^d), \quad g \cdot \nu \equiv 0 \text{ in } \Sigma,$$

identified explicitly in the proof below, see (3.5.27), such that, once inserted in (the Dirichlet B.C. (3.1.2d) of) the linearized problem (3.1.2) or (3.1.4a), yields a solution y with the following properties, for all $\epsilon > 0$:

$$(3.5.2) \qquad y^0 \in \left(H^{\frac{1}{2}+\epsilon}(\Omega)\right)^d \cap H$$
$$\Rightarrow y \in L^2\left(0,\infty;\left(H^{\frac{3}{2}+\epsilon}(\Omega)\right)^d \cap H\right) \cap H^{\frac{3}{4}+\frac{\epsilon}{2}}(0,\infty;H).$$

*This way, the Finite Cost Condition (3.1.22)–(3.1.25) for problem (3.1.4a) is satisfied for $d = 3$. The successful boundary control g given by (3.5.21) below is the restriction on $\Gamma = \partial\Omega$ of the solution of the interior stabilization problem provided by [**B-T.1**] for the linearized problem over a larger domain $\tilde{\Omega}$, as described below.*

PROOF. STEP 1. We first extend the original domain Ω into a slightly larger domain $\tilde{\Omega} = \Omega \cup \omega$, where ω is an arbitrarily small open set across all of Γ from Ω, so that ω and Ω share the common boundary $\Gamma : \partial\omega \cap \partial\Omega = \Gamma$. Next, we extend the initial condition y_0 (in this proof we call the I.C. y_0 rather than y^0 for convenience) and the pressure p from Ω to $\tilde{\Omega}$, in the usual way [**Te.1**, Appendix A]. To this end, for the purpose of this proof, we shall note explicitly the dependence on the relevant domain. Thus, recalling (1.5a), we shall write

(3.5.3a) $\quad H \equiv H_\Omega = \{y \in (L^2(\Omega))^d : \nabla \cdot y \equiv 0 \text{ in } \Omega; \ y \cdot \nu \equiv 0 \text{ in } \partial\Omega\};$

(3.5.3b) $\quad \tilde{H} \equiv H_{\tilde{\Omega}} = \{y \in (L^2(\tilde{\Omega}))^d : \nabla \cdot y \equiv 0 \text{ in } \tilde{\Omega}; \ y \cdot \nu \equiv 0 \text{ in } \partial\tilde{\Omega}\}.$

Henceforth, we shall take Ω simply connected, as assumed.

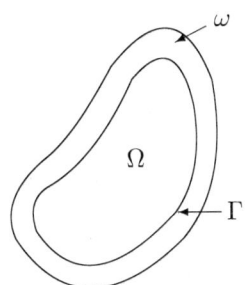

FIG. 3.5.1:
y_0 NON-VANISHING ON Γ

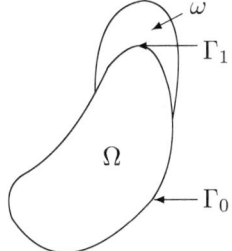

FIG. 3.5.2:
y_0 VANISHING ON Γ_0

Step 2. (Extension procedure, $d = 3$) (i) Let $f \in H \equiv H_\Omega$ [but $f \in (L^2(\Omega))^3$, with $\nabla \cdot f \equiv 0$ in Ω, $\int_{\Gamma_i} f \cdot \nu \, d\Gamma = 0$ will suffice Γ_i being the finitely many boundary connected components]. By [**Te.1**, Prop. 1.3, p. 467], [**Ce.1**, Thm. 9, p. 54], there exists $F \in (H^1(\Omega))^3$ such that: curl $F = f$ in Ω. Moreover, by [**Te.1**, Lemmas 1.4 and 1.5, pp. 464–5], [**Ce.1**, Thm. 9, p. 54], we may further require that—for Ω connected as assumed—F satisfies the additional condition $F \in H \equiv H_\Omega$. [Moreover, F is unique [**Te.1**, Remark 1.1(i), p. 463] for Ω simply connected. Otherwise, uniqueness holds true modulo a finite-dimensional subspace, of dimension equal to the number of 'handles' [Giles Auchmuty, private communication]. See also [**Au.1**, Eqn. (3.8)].] In conclusion: given $f \in H \equiv H_\Omega$, there exists an $F \in (H^1(\Omega))^3 \cap H$, such that

(3.5.4) $\quad$ curl $F = f \in H_\Omega; \quad \nabla \cdot F \equiv 0 \text{ in } \Omega; \quad F \cdot \nu = 0 \text{ on } \partial\Omega.$

(ii) Next, let $f \in (H^1(\Omega))^3 \cap H$. By part (i), there exists *a-fortiori* a unique $F \in (H^1(\Omega))^3 \cap H$ such that: curl $F = f$ in Ω. Actually, by [**Te.1**, Prop. 1.4, $m = 1$, p. 457], we obtain that $F \in (H^2(\Omega))^3 \cap H$, so that

(3.5.5) $\quad$ curl $F = f \in (H^1(\Omega))^3 \cap H; \quad \nabla \cdot F \equiv 0 \text{ in } \Omega; \quad F \cdot \nu = 0 \text{ on } \partial\Omega.$

(iii) By interpolation between part (i), Eqn. (3.5.4), and part (ii), Eqn. (3.5.5), if $f \in (H^s(\Omega))^3 \cap H$, $0 \leq s \leq 1$, there exists a unique $F \in (H^{s+1}(\Omega))^3 \cap H$, such that

(3.5.6) $\quad$ curl $F = f \in (H^s(\Omega))^3 \cap H; \quad \nabla \cdot F \equiv 0 \text{ in } \Omega; \quad F \cdot \nu = 0 \text{ on } \partial\Omega.$

(iv) Next, define $\tilde{\Omega} = \Omega \cup \omega$, where ω is an open strip, exterior to Ω, around all of $\Gamma = \partial\Omega$. Let $f \in (H^s(\Omega))^3 \cap H$, $\frac{1}{2} < s \leq 1$, and $F \in (H^{s+1}(\Omega))^3 \cap H$, corresponding to part (iii), Eqn. (3.5.6). By usual symmetry techniques, we extend such F across the boundary Γ onto $\tilde{\Omega}$ to obtain $\tilde{F} \in (H^{s+1}(\tilde{\Omega}))^3$. We then define $\hat{F} \equiv \rho \tilde{F}$, where $\rho \in C_0^\infty(\tilde{\Omega})$, $\rho \equiv 1$ on $\overline{\Omega}$. Finally, we define

(3.5.7) $\quad\begin{cases} \tilde{f} \equiv \operatorname{curl} \hat{F} \in (H^s(\tilde{\Omega}))^3, \text{ so that } \nabla \cdot \tilde{f} \equiv 0 \text{ in } \tilde{\Omega};\ \tilde{f} \equiv 0 \text{ near } \partial\tilde{\Omega}; \\ \tilde{f} \equiv f \text{ in } \Omega, \text{ and we conclude that } \tilde{f} \in (H_0^s(\tilde{\Omega}))^3 \cap \tilde{H}. \end{cases}$

(v) We apply the extension procedure described in part (iv) from f in Ω to $\tilde{f}$ in $\tilde{\Omega}$ to the initial condition y_0 for $s = \frac{1}{2} + \epsilon$. Thus:

(3.5.8) given $y_0 \in (H^{\frac{1}{2}+\epsilon}(\Omega))^3 \cap H$, we can extend it to $\tilde{y}_0 \in (H_0^{\frac{1}{2}+\epsilon}(\tilde{\Omega}))^3 \cap \tilde{H}$.

REMARK 3.5.1. (a) Generally, for y_0 non-vanishing on any part of $\Gamma = \partial\Omega$, we take ω to be a strip all around Γ, and then $\tilde{\Omega} = \Omega \cup \omega$. See Fig. 3.5.1.

(b) If, however, y_0 vanishes on the portion Γ_0 of Γ, we may take ω to be an open set, exterior to Ω, just across the complementary part $\Gamma_1 = \Gamma \setminus \Gamma_0$, and then $\tilde{\Omega} = \Omega \cup \omega$. See Fig. 3.5.2.

(c) If y_0 vanishes on all of Γ, we may take ω to be an *arbitrarily small* open set, exterior to Ω just across an *arbitrarily small* portion Γ_1 of Γ, and then $\tilde{\Omega} = \Omega \cup \omega$. See Fig. 3.5.2 with Γ_1 arbitrarily small.

In each of the three cases above, we obtain that the extension $\tilde{y}_0$ of the original Initial Condition $y_0 \in (H^{\frac{1}{2}+\epsilon}(\Omega))^3 \cap H$ satisfies the property: $\tilde{y}_0 \in (H_0^{\frac{1}{2}+\epsilon}(\tilde{\Omega}))^3 \cap \tilde{H}$.

(vi) Given the steady state solution $y_e \in (H^2(\Omega))^3 \cap V$, we can extend it through the above procedure as given in (iv) to $\tilde{y}_e \in (H^2(\tilde{\Omega}))^3 \cap \tilde{V}$, by invoking [**Te.1**, Prop. 1.4, for $m = 2$, p. 467], where $\tilde{V} = \{h \in (H_0^1(\tilde{\Omega}))^3;\ \nabla \cdot h \equiv 0 \text{ in } \tilde{\Omega}\}$. Likewise given the pressure $p \in H^1(\Omega)$, we can extend it to $\tilde{p} \in H^1(\tilde{\Omega})$. We have thus extended y_0, y_e, p in Ω to $\tilde{y}_0$, $\tilde{y}_e$, $\tilde{p}$ in $\tilde{\Omega}$.

Step 3. (Interior feedback stabilization on $\tilde{\Omega} = \Omega \cup \omega$) With $\tilde{y}_0 \in (H_0^{\frac{1}{2}+\epsilon}(\Omega))^3 \cap \tilde{H}$, $\tilde{H} \equiv H_{\tilde{\Omega}}$, an extension of the original initial condition for the linearized problem (3.1.4) over Ω, constructed as in Steps 1,2, above, we now invoke the interior feedback stabilization results of [**B-T.1**] over all of $\tilde{\Omega}$, with finite-dimensional interior controller supported on the auxiliary set ω. More precisely, there exists a s.c. analytic (feedback) semigroup $e^{\tilde{A}_F t}$, defined on the space $\tilde{H} \equiv H_{\tilde{\Omega}}$ in (3.5.3b), such that the function

(3.5.9) $\quad \tilde{y}(t; \tilde{y}_0) = e^{\tilde{A}_F t} \tilde{y}_0, \quad t \geq 0, \quad \tilde{A}_F = \tilde{A} + \tilde{P}\tilde{F}_\omega, \quad \mathcal{D}(\tilde{A}_F) = \mathcal{D}(\tilde{A}) = \mathcal{D}(\tilde{\mathcal{A}})$

is the (unique) solution of the following problem

(3.5.10a) $\quad\begin{cases} \tilde{y}_t - \nu_0 \Delta \tilde{y} + (\tilde{y}_e \cdot \nabla)\tilde{y} + (\tilde{y} \cdot \nabla)\tilde{y}_e = \tilde{F}_\omega(\tilde{y}) + \nabla \tilde{p} & \text{in } (0, \infty) \times \tilde{\Omega}; \\ \nabla \cdot \tilde{y} = 0 & \text{in } (0, \infty) \times \tilde{\Omega}; \\ \tilde{y}|_{\tilde{\Gamma}} = 0 & \text{in } (0, \infty) \times \tilde{\Gamma},\ \tilde{\Gamma} = \partial \tilde{\Omega}; \\ \tilde{y}(0, \cdot) = \tilde{y}_0 & \text{in } \tilde{\Omega}. \end{cases}$

(3.5.10b)

(3.5.10c)

(3.5.10d)

Here $\tilde{P}$ is the Leray orthogonal projector $L^2(\tilde{\Omega}) \to H_{\tilde{\Omega}}$, and $\tilde{F}_\omega$ is a finite-dimensional linear operator acting only on the restriction $\tilde{y}|_\omega$ of $\tilde{y}$ on ω. Reference [**B-T.1**,

Theorem 2.1, Section 2; Lemma 3.1, Section 3.2] provides two different specific expressions for $\tilde{F}_\omega$, both covered by the common form

$$(3.5.10e) \qquad \tilde{F}_\omega(\tilde{y}) = m\left(-\sum_{i=1}^{2K}(\tilde{y}, \tilde{a}_i)_\omega \tilde{b}_i\right)$$

for suitable vectors $\tilde{a}_i$, $\tilde{b}_i$ (whichever form we use is irrelevant to the present proof), where m is the characteristic function of the auxiliary set ω: $m \equiv 1$ on ω, $m = 0$ otherwise. Finally, $\tilde{y}_e$ and $\tilde{p}$ are extensions on $\tilde{\Omega}$ of the original y_e and p on Ω, constructed as in Steps 1,2. A sign $\sim$ on top of a symbol denotes that that quantity refers to the augmented domain $\tilde{\Omega}$.

Moreover, the s.c. semigroup $e^{\tilde{A}_F t}$ is uniformly stable on $\tilde{H} \equiv H_{\tilde{\Omega}}$: there exist constants $\tilde{c}$, $\tilde{\gamma} > 0$, such that [**B-T.1**]

$$(3.5.11) \qquad |\tilde{y}(t; \tilde{y}_0)|_{\tilde{H}} = \left|e^{\tilde{A}_F t}\tilde{y}_0\right|_{\tilde{H}} \leq \tilde{c}e^{-\tilde{\gamma}t}|\tilde{y}_0|_{\tilde{H}}, \quad t \geq 0.$$

Moreover, $\tilde{A}_F$ is a bounded perturbation of $\tilde{\mathcal{A}}$ on $\tilde{H}$, at least under [**B-T.1**, Lemma 3.1, Section 3.2]. Under the Riccati-based feedback as in [**B-T.1**, Theorem 2.1], the Riccati operator R_N satisfies $|R_N y|_H \leq c\|y\|$ and hence acts topologically like $\mathcal{A}_0$. At any rate, the domains of fractional powers $\mathcal{D}((-\tilde{A}_F)^\theta)$ of $(-\tilde{A}_F)$ coincide with the domains of fractional powers of a suitable translation $(\mu I - \tilde{\mathcal{A}})^\theta$ of $(-\mathcal{A})$ for a positive $\mu > 0$, which in turn coincide with the domains of fractional powers of the positive, self-adjoint operator $\tilde{A}^\theta$ on $\tilde{H} \equiv H_{\tilde{\Omega}}$ (see (1.13)–(1.17)):

$$(3.5.12) \qquad \mathcal{D}((-\tilde{A}_F)^\theta) = \mathcal{D}((\mu I - \tilde{\mathcal{A}})^\theta) = \mathcal{D}(\tilde{A}^\theta), \quad 0 \leq \theta \leq 1.$$

We next recall (3.5.8); in light of (3.5.12) and (1.1.5), we now complement it as follows:

Given $y_0 \in (H^{\frac{1}{2}+\epsilon}(\Omega))^3 \cap H$, we can extend it to

$$(3.5.13) \qquad \tilde{y}_0 \in \left(H_0^{\frac{1}{2}+\epsilon}(\tilde{\Omega})\right)^3 \cap \tilde{H} \equiv \mathcal{D}\left(\tilde{A}^{\frac{1}{4}+\frac{\epsilon}{2}}\right) \equiv \mathcal{D}\left((-\tilde{A}_F)^{\frac{1}{4}+\frac{\epsilon}{2}}\right).$$

Then, we first can complement (3.5.11) as follows, for $\tilde{y}_0$ as in (3.5.13)

$$(3.5.14) \quad |\tilde{y}(t;\tilde{y}_0)|_{(H_0^{\frac{1}{2}+\epsilon}(\tilde{\Omega}))^3\cap\tilde{H}} = \left|e^{\tilde{A}_F t}\tilde{y}_0\right|_{\mathcal{D}((-\tilde{A}_F)^{\frac{1}{4}+\frac{\epsilon}{2}})}$$

$$= \left|e^{\tilde{A}_F t}(-\tilde{A}_F)^{\frac{1}{4}+\frac{\epsilon}{2}}\tilde{y}_0\right|_{\tilde{H}} \leq \tilde{c}e^{-\tilde{\gamma}t}\left|(-\tilde{A}_F)^{\frac{1}{4}+\frac{\epsilon}{2}}\tilde{y}_0\right|_{\tilde{H}}$$

$$= \tilde{c}e^{-\tilde{\gamma}t}|\tilde{y}_0|_{(H_0^{\frac{1}{2}+\epsilon}(\tilde{\Omega}))^3\cap\tilde{H}}, \quad t \geq 0.$$

Regularity of $\tilde{y}$. We now establish the following critical regularity of $\tilde{y}(t;\tilde{y}_0) = e^{\tilde{A}_F t}\tilde{y}_0$ in (3.5.9):

$$(3.5.15) \quad \begin{cases} \tilde{y}_0 \in (H_0^{\frac{1}{2}+\epsilon}(\tilde{\Omega}))^3 \cap \tilde{H} \equiv \mathcal{D}((-\tilde{A}_F)^{\frac{1}{4}+\frac{\epsilon}{2}}) \\ \Rightarrow \tilde{y} \in L^2(0,\infty; [(H^{\frac{3}{2}+\epsilon}(\tilde{\Omega}))^3 \cap \tilde{V}]) \cap H^{\frac{3}{4}+\frac{\epsilon}{2}}(0,\infty; \tilde{H}). \end{cases}$$

In fact, using (3.5.12), (3.5.13), (3.5.14), we first obtain the following regularity result

$$(3.5.16) \begin{cases} \tilde{y}_0 \in \left(H_0^{\frac{1}{2}+\epsilon}(\tilde{\Omega})\right)^3 \cap \tilde{H} \equiv \mathcal{D}\left((-\tilde{A}_F)^{\frac{1}{4}+\frac{\epsilon}{2}}\right) \\ \Rightarrow \tilde{y}(t;\tilde{y}_0) = e^{\tilde{A}_F t}\tilde{y}_0 = (-\tilde{A}_F)^{-(\frac{1}{4}+\frac{\epsilon}{2})}e^{\tilde{A}_F t}(-\tilde{A}_F)^{\frac{1}{4}+\frac{\epsilon}{2}}\tilde{y}_0 \\ \qquad \in L^2\left(0,\infty; \mathcal{D}\left((-\tilde{A}_F)^{\frac{3}{4}+\frac{\epsilon}{2}}\right) \equiv \left(H^{\frac{3}{2}+\epsilon}(\tilde{\Omega})\right)^3 \cap \tilde{V}\right) \\ \qquad \cap \ C\left([0,T]; \mathcal{D}\left((-\tilde{A}_F)^{\frac{1}{4}+\frac{\epsilon}{2}}\right) \equiv \left(H_0^{\frac{1}{2}+\epsilon}(\tilde{\Omega})\right)^3 \cap \tilde{H}\right), \end{cases}$$

recalling (1.15). Thus, the interior control leads to smooth solutions compatible with domains of fractional powers of A. Next, from (3.5.16) and from Eqn. (3.5.10a) acted upon by the Leray projector $\tilde{P}$ we have, since $\tilde{P}\tilde{y}_t = \tilde{y}_t$ as $\tilde{y} \in \tilde{H}$:

$$(3.5.17) \quad \begin{aligned} \tilde{P}\tilde{y}_t(t;\tilde{y}_0) = \tilde{y}_t(t;\tilde{y}_0) &= \tilde{P}[\nu_0 \Delta \tilde{y} - (\tilde{y}_e \cdot \nabla)\tilde{y} - (\tilde{y} \cdot \nabla)\tilde{y}_e + \tilde{F}_\omega(\tilde{y})] \\ &\in L^2\left(0,\infty; (H^{-\frac{1}{2}+\epsilon}(\tilde{\Omega}))\right). \end{aligned}$$

Therefore, by (3.5.16) and (3.5.17), using the intermediate derivative theorem [**L-M.1**, Vol. 1, p. 15] (in fractional form), we deduce for the time derivative D_t^θ of order θ that

$$(3.5.18) \quad D_t^\theta \tilde{y} \in L^2(0,\infty; (H^{\frac{3}{2}-2\theta+\epsilon}(\tilde{\Omega}))^d), \ 0 \leq \theta \leq 1,$$

$$[H^{\frac{3}{2}+\epsilon}, H^{-\frac{1}{2}+\epsilon}]_\theta \equiv H^{\frac{3}{2}-2\theta+\epsilon}.$$

In particular, we have $\frac{3}{2} - 2\theta + \epsilon = 0$ for $\theta = \frac{3}{4} + \frac{\epsilon}{2}$, in which case (3.5.18) specializes to

$$(3.5.19) \quad D_t^{\frac{3}{4}+\frac{\epsilon}{2}}\tilde{y} \in L^2(0,\infty; H), \text{ or } \tilde{y} \in H^{\frac{3}{4}+\frac{\epsilon}{2}}(0,\infty; H).$$

Then (3.5.16) and (3.5.19) establish (3.5.15).

Regularity and exponential decay of the trace $\tilde{y}(t;\tilde{y}_0)|_\Gamma$ **on** $\Gamma \equiv \partial \Omega$. To get the regularity of $\tilde{y}|_\Gamma$, we apply the trace theorem (in both space and time) as in [**L-M.1**, Vol. 2, p. 9] to (3.5.15) and obtain

$$(3.5.20) \begin{cases} \tilde{y}_0 \in (H_0^{\frac{1}{2}+\epsilon}(\tilde{\Omega}))^3 \cap \tilde{H} \equiv \mathcal{D}((-\tilde{A}_F)^{\frac{1}{4}+\frac{\epsilon}{2}}) \\ \Rightarrow \tilde{y}(t;\tilde{y}_0)|_\Gamma \in H^{1+\epsilon,\frac{1}{2}+\frac{\epsilon}{2}}(\Sigma_\infty) \\ \qquad \equiv L^2(0,\infty; (H^{1+\epsilon}(\Gamma))^3) \cap H^{\frac{1}{2}+\frac{\epsilon}{2}}(0,\infty; (L^2(\Gamma))^3). \end{cases}$$

We now obtain *exponential decay* of $\tilde{y}|_\Gamma$. For y_0 and $\tilde{y}_0$ as in (3.5.8), (3.5.13), the above well-posed problem (3.5.10): $\tilde{y}(t;\tilde{y}_0) = e^{\tilde{A}_F t}\tilde{y}_0$ on $\tilde{\Omega}$ produces via (3.5.14), (3.5.15), or (3.5.20), a well-defined trace or restriction over the boundary $\Gamma = \partial \Omega$ of the original domain Ω satisfying:

$$(3.5.21) \quad |\tilde{y}(t;\tilde{y}_0)|_\Gamma|_{(L^2(\Gamma))^3} = \left|e^{\tilde{A}_F t}\tilde{y}_0|_\Gamma\right|_{(L^2(\Gamma))^3} \leq \tilde{c}e^{-\tilde{\gamma}t}|\tilde{y}_0|_{(H_0^{\frac{1}{2}+\epsilon}(\tilde{\Omega}))^3 \cap \tilde{H}}, \ t \geq 0.$$

Step 4. We next consider the restriction $\tilde{y}(t;\tilde{y}_0)|_\Omega$ of the solution $\tilde{y}(t;\tilde{y}_0)$ of problem (3.5.10) over the original smaller domain Ω by setting:

$$(3.5.22) \quad \hat{y}(t;\tilde{y}_0) \stackrel{\text{def}}{\equiv} \chi\tilde{y}(t;\tilde{y}_0) = \tilde{y}(t;\tilde{y}_0)|_\Omega = e^{\tilde{A}_F t}\tilde{y}_0|_\Omega, \; t \geq 0,$$

$$\in \; L^2(0,\infty; [(H^{\frac{3}{2}+\epsilon}(\Omega))^3 \cap H]) \cap H^{\frac{3}{4}+\frac{\epsilon}{2}}(0,\infty;H),$$

by (3.5.15), where χ is the characteristic function of the set Ω: $\chi \equiv 1$ on Ω; and $\chi \equiv 0$ elsewhere. Then, considering problem (3.5.10) only on Ω—or multiplying problem (3.5.10) by χ—we see that the function $\hat{y}(t;\tilde{y}_0)$ defined in (3.5.16) satisfies the following problem:

$$
\begin{cases}
(3.5.23\text{a}) & \hat{y}_t - \nu_0 \Delta \hat{y} + (y_e \cdot \nabla)\hat{y} + (\hat{y} \cdot \nabla)y_e = \nabla p & \text{in } (0,\infty) \times \Omega; \\
(3.5.23\text{b}) & \nabla \cdot \hat{y} = 0 & \text{in } (0,\infty) \times \Omega; \\
(3.5.23\text{c}) & \hat{y}|_\Gamma = \tilde{y}(t;\tilde{y}_0)|_\Gamma & \text{in } (0,\infty) \times \Gamma; \\
(3.5.23\text{d}) & \hat{y}(0,\cdot) = \tilde{y}(0;\tilde{y}_0)|_\Omega = \tilde{y}_0|_\Omega = y_0 & \text{in } \Omega,
\end{cases}
$$

since $p = \tilde{p}|_\Omega$; $y_e = \tilde{y}_e|_\Omega$, $y_0 = \tilde{y}_0|_\Omega$ for the restrictions of $\tilde{p}$, $\tilde{y}_e$, and $\tilde{y}_0$ on Ω according to the procedure of Step 1, while $\chi \tilde{F}_\omega(\tilde{y}) = 0$, since—referring to (3.5.10e)—we have $\chi m(\tilde{b}_i) = 0$. This is so since χ vanishes on ω (which is external to Ω), where $m\tilde{b}_i = \tilde{b}_i$, while $m\tilde{b}_i = 0$ in Ω. We also have $\int_\Gamma \tilde{y} \cdot \nu \, d\Gamma = 0$ by (3.5.10b). From (3.5.22) and (3.5.11), we obtain that $\hat{y}(t;\tilde{y}_0)$ decays with exponential rate on $H = H_\Omega$:

$$(3.5.24) \quad |\hat{y}(t;\tilde{y}_0)|_{(L^2(\Omega))^3} \leq c e^{-\tilde{\gamma} t} |\tilde{y}_0|_H, \quad t \geq 0.$$

The fact that we put a "hat" $\hat{}$ on the solution of problem (3.5.23) indicates that this is only an intermediary or auxiliary problem. Indeed, it has the deficiency that $\hat{y}|_\Gamma \cdot \nu \neq 0$ on Σ, in general.

For future reference, we now project problem (3.5.23) onto $H \equiv H_\Omega$. Since generally $\hat{y}|_\Gamma \cdot \nu \neq 0$ on Σ, we have that $D[\tilde{y}(\,\cdot\,;\tilde{y}_0)|_\Gamma] \notin H \equiv H_\Omega$ generally, hence the projection of (3.5.23) onto H may be written as the abstract equation (recalling (3.1.4b) with $\int_\Gamma \tilde{y} \cdot \nu \, d\Gamma = 0$, as observed above

$$(3.5.25\text{a}) \quad (P\hat{y})' - \mathcal{A}(P\hat{y}) = -\mathcal{A}PD[\tilde{y}(\,\cdot\,;\tilde{y}_0)|_\Gamma], \quad y(0) = y_0 \in H;$$

$$(3.5.25\text{b}) \quad P\hat{y}(t;y_0) = e^{\mathcal{A}t}y_0 - \int_0^t e^{\mathcal{A}(t-\tau)}\mathcal{A}PD[\tilde{y}(\tau;\tilde{y}_0)|_\Gamma]d\tau, \; Py_0 = y_0.$$

Step 5. We consider the following problem on Ω:

$$
\begin{cases}
(3.5.26\text{a}) & y_t - \nu_0 \Delta y + (y_e \cdot \nabla)y + (y \cdot \nabla)y_e = \nabla p & \text{in } (0,\infty) \times \Omega; \\
(3.5.26\text{b}) & \nabla \cdot y = 0 & \text{in } (0,\infty) \times \Omega; \\
(3.5.26\text{c}) & y|_\Gamma = g \equiv \tilde{y}(t;\tilde{y}_0)|_\Gamma - (\tilde{y}(t;\tilde{y}_0)|_\Gamma \cdot \nu)\nu & \text{in } (0,\infty) \times \Gamma; \\
(3.5.26\text{d}) & y(0,\cdot) = y_0 & \text{in } \Omega.
\end{cases}
$$

We notice that the y-problem in (3.5.26) differs from the $\hat{y}$-problem in (3.5.23) simply in the fact that, on the boundary Γ, we have subtracted in (3.5.26c) the normal component $(\tilde{y}(t; \tilde{y}_0)|_\Gamma \cdot \nu)\nu$ of the boundary term $\tilde{y}(t; \tilde{y}_0)|_\Gamma$ in (3.5.23c). Thus, the boundary control

(3.5.27) $\qquad g(t; \tilde{y}_0) \equiv \tilde{y}(t; \tilde{y}_0)|_\Gamma - (\tilde{y}(t; \tilde{y}_0)|_\Gamma \cdot \nu)\nu \quad \text{on } (0, \infty) \times \Gamma$

is a tangential vector on Γ, so that it satisfies the required pointwise boundary compatibility condition ((3.1.2e) left)

(3.5.28) $\qquad\qquad g(t; \tilde{y}_0) \cdot \nu \equiv 0 \quad \text{on } (0, \infty) \times \Gamma.$

Moreover, from (3.5.27) and (3.5.21), we obtain (recall (1.15))

(3.5.29) $\qquad |g(t; \tilde{y}_0)|_{(L^2(\Gamma))^3} \leq \tilde{c} e^{-\tilde{\gamma} t} |\tilde{y}_0|_{\mathcal{D}(\tilde{A}^{\frac{1}{4} + \frac{\epsilon}{2}})},$

$$\mathcal{D}\left(\tilde{A}^{\frac{1}{4} + \frac{\epsilon}{2}}\right) \equiv (H_0^{\frac{1}{2}+\epsilon}(\tilde{\Omega}))^3 \cap \tilde{H}, \ t \geq 0.$$

Finally, recalling the regularity (3.5.20) of the trace $\tilde{y}|_\Gamma$ as well as the definition (3.5.27) of g, we obtain

(3.5.30) $\qquad \begin{cases} \tilde{y}_0 \in (H_0^{\frac{1}{2}+\epsilon}(\tilde{\Omega}))^3 \cap \tilde{H} \equiv \mathcal{D}((-\tilde{A}_F)^{\frac{1}{4}+\frac{\epsilon}{2}}) \\ \Rightarrow g \in H^{1+\epsilon, \frac{1}{2}+\frac{\epsilon}{2}}(\Sigma_\infty) \\ \equiv L^2(0, \infty; (H^{1+\epsilon}(\Gamma))^3) \cap H^{\frac{1}{2}+\frac{\epsilon}{2}}(0, \infty; (L^2(\Gamma))^3). \end{cases}$

$\square$

Our next goal is to establish that the solution $y(t; y_0)$ of (3.5.26) decays exponentially in time in $H \equiv H_\Omega$. Note that the difficulty is due to the generator being unstable.

PROPOSITION 3.5.2. *With reference to problem (3.5.26), we have that $y(t; \tilde{y}_0)$ is exponentially stable in the following sense: there exists $C > 0$ such that*

(3.5.31) $\qquad |y(t; \tilde{y}_0)|_H \leq C e^{-(\gamma_0 - \epsilon)t} |\tilde{y}_0|_{\mathcal{D}(\tilde{A}^{\frac{1}{4}+\frac{\epsilon}{2}})}, \ t \geq 0,$

where $0 < \gamma < \tilde{\gamma}$, where $\tilde{\gamma}$ is defined by (3.5.29).

REMARK 3.5.2. In order to study the exponential stability of the y-problem in (3.5.26), one might at first be tempted to consider the term $-(\tilde{y}(t; \tilde{y}_0)|_\Gamma \cdot \nu)\nu$ in (3.5.26c) as a boundary non-homogeneous term for the $\hat{y}$-"free dynamics" (3.5.23), since such solution $\hat{y}(t; \tilde{y}_0)$ has the attraction of being exponentially stable to begin with, see (3.5.24). Likewise, such non-homogeneous boundary term $(\tilde{y}(t; \tilde{y}_0)|_\Gamma \cdot \nu)\nu$ is also exponentially stable to begin with, see (3.5.21). The problem with this approach, however, is that the "free dynamics" (3.5.23), once projected onto $H \equiv H_\Omega$ by the Leray orthogonal projection P, is not a s.c. semigroup on H. Thus, a variation parameter formula (of the type in (3.5.32) below) to account for the non-homogeneous boundary term $-(\tilde{y}(t; \tilde{y}_0)|_\Gamma \cdot \nu)\nu$ is not available.

Therefore, we resort to a different approach: we consider the entire exponentially stable (see (3.5.29)) boundary term $g(t; \tilde{y}_0)$ in (3.5.27) as a non-homogeneous term for the y-problem (3.5.26) which—once projected by the Leray projector P onto $H \equiv H_\Omega$—is driven by the semigroup $e^{\mathcal{A}t}$, even though $e^{\mathcal{A}t}$ is unstable. Indeed, if we set $g \equiv 0$ in the B.C. (3.5.26c), see (3.5.27), we obtain the well-posed system $y(t; y_0) = e^{\mathcal{A}t} y_0$ on $H = H_\Omega$.

PROOF OF PROPOSITION 3.5.2. *Step (i)*. Following the second approach described in Remark 3.5.2, we rewrite the y-problem (3.5.26) on $H \equiv H_\Omega$ abstractly—by virtue of the key pointwise boundary c.c. (3.5.28)—by recalling Eqn. (3.1.4a). Thus, we obtain

(3.5.32a) $$y_t - \mathcal{A}y = -\mathcal{A}Dg, \qquad y(0) = y_0 \in H \equiv H_\Omega,$$

with g defined by (3.5.27). The corresponding variation of parameter formula is

(3.5.32b) $$y(t) = e^{\mathcal{A}t}y_0 - \int_0^t e^{\mathcal{A}(t-\tau)}\mathcal{A}Dg(\tau;\tilde{y}_0)d\tau$$

(3.5.32c) $$= e^{\mathcal{A}t}y_0 - \int_0^t e^{\mathcal{A}(t-\tau)}\mathcal{A}D[\tilde{y}(\tau;\tilde{y}_0)|_\Gamma - (\tilde{y}(\tau;\tilde{y}_0)|_\Gamma \cdot \nu)\nu]d\tau.$$

Step (ii). We return to the dynamics for $P\hat{y}(t;y_0)$ in (3.5.25b) and rewrite it in terms of its stable and unstable components, see (3.4.10), (3.4.9). Recalling $P_N, \mathcal{A}_N^u, \mathcal{A}_N^s, \ldots$ from (3.4.3), (3.4.5), setting $y_0^u = P_N y_0$, $y_0^s = (I - P_N)y_0$, we obtain from (3.5.25b):

(3.5.33) $$\begin{cases} P\hat{y}(t;y_0) = e^{\mathcal{A}_N^u t}y_0^u - \int_0^t e^{\mathcal{A}_N^u(t-\tau)}P_N(\mathcal{A}PD)[\tilde{y}(\tau;\tilde{y}_0)|_\Gamma]d\tau + \eta(t;y_0) \\ \\ \eta(t;y_0) \equiv e^{\mathcal{A}_N^s t}y_0^s - \int_0^t e^{\mathcal{A}_N^s(t-\tau)}(I-P_N)(\mathcal{A}PD)[\tilde{y}(\tau;\tilde{y}_0)|_\Gamma]d\tau \end{cases}$$

(3.5.34) $$\qquad = e^{\mathcal{A}_N^s t}y_0^s - \int_0^t (-\mathcal{A}_N^s)^{\frac{3}{4}+\epsilon}e^{\mathcal{A}_N^s(t-\tau)}(-\mathcal{A}_N^s)^{\frac{1}{4}-\epsilon} \\ \cdot (I-P_N)PD[\tilde{y}(\tau;\tilde{y}_0)|_\Gamma]d\tau.$$

By (3.4.7), (3.1.3e), and (3.5.21) used in (3.5.34), we obtain,

(3.5.35) $$|\eta(t;y_0)|_H \leq C_{\gamma_0}e^{-\gamma_0 t}|y_0^s|_H$$
$$+ C_{\gamma_0}\int_0^t \frac{e^{-\gamma_0 t}e^{-(\tilde{\gamma}-\gamma_0)\tau}}{(t-\tau)^{\frac{3}{4}+\epsilon}}d\tau|\tilde{y}_0|_{\mathcal{D}(\tilde{A}^{\frac{1}{4}+\frac{\epsilon}{2}})}$$
$$\leq Ce^{-\gamma_0 t}|y_0^s|_H + Ce^{-\gamma_0 t}t^{\frac{1}{4}-\epsilon}|\tilde{y}_0|_{\mathcal{D}(\tilde{A}^{\frac{1}{4}+\frac{\epsilon}{2}})}$$
$$\leq Ce^{-(\gamma_0-\epsilon)t}|\tilde{y}_0|_{\mathcal{D}(\tilde{A}^{\frac{1}{4}+\frac{\epsilon}{2}})},$$

taking $0 < \gamma_0 < \tilde{\gamma}$, which is possible, since γ_0 in (3.4.7) is any number $0 < \gamma_0 < |\text{Re }\lambda_{N+1}|$.

Step (iii). A similar procedure applied to the y-dynamics (3.5.32c) yields

(3.5.36a) $$\begin{cases} y(t;y_0) = e^{\mathcal{A}_N^u t}y_0^u - \int_0^t e^{\mathcal{A}_N^u(t-\tau)}P_N(\mathcal{A}D)g(\tau;\tilde{y}_0)d\tau + \zeta(t;y_0) \\ \\ \zeta(t;y_0) = e^{\mathcal{A}_N^s t}y_0^s - \int_0^t (-\mathcal{A}_N^s)^{\frac{3}{4}+\epsilon}e^{\mathcal{A}_N^s(t-\tau)}(-\mathcal{A}_N^s)^{\frac{1}{4}-\epsilon} \\ \qquad\qquad\qquad \cdot (I-P_N)Dg(\tau;\tilde{y}_0)d\tau, \end{cases}$$

(3.5.36b)

where taking again $0 < \gamma_0 < \tilde{\gamma}$, which is possible, we obtain by the decay of g in (3.5.29)

$$|\zeta(t; y_0)|_H \leq Ce^{-(\gamma_0-\epsilon)t}|\tilde{y}_0|_{\mathcal{D}(\tilde{\mathcal{A}}^{\frac{1}{4}+\frac{\epsilon}{2}})}, \quad t \geq 0. \tag{3.5.37}$$

Step (iv). To the dynamics (3.5.33) for $P\hat{y}(t; y_0)$ and (3.5.36) for $y(t; y_0)$ we shall apply the following (standard) result [**M-M.1**, p. 268]. $\square$

LEMMA 3.5.3. *Consider the (finite-dimensional) dynamics*

$$r(t) = e^{\mathcal{A}_N^u t} y_0^u + \int_0^t e^{\mathcal{A}_N^u(t-\tau)} P_N(\mathcal{A}PD)f(\tau)d\tau, \; t \geq 0, \tag{3.5.38}$$

in $H \equiv H_\Omega$, where the boundary datum f satisfies

$$|f(t)|_{(L^2(\Gamma))^3} \leq Ce^{-at}, \; a > 0, \; t \geq 0; \quad \int_\Gamma f \cdot \nu \, d\Gamma = 0. \tag{3.5.39}$$

Then, a necessary and sufficient condition to have $r(t) \to 0$ as $t \to \infty$, is that the condition

$$y_0^u + \int_0^\infty e^{(-\mathcal{A}_N^u)\tau} P_N(\mathcal{A}PD)f(\tau)d\tau = 0 \tag{3.5.40}$$

holds true, in which case, in fact, we have

$$|r(t)|_{L^2(\Omega))^3} \leq Ce^{-at}, \; t \geq 0. \tag{3.5.41}$$

PROOF LEMMA 3.5.3. We rewrite (3.5.38) as follows:

$$\begin{cases} r(t) = e^{\mathcal{A}_N^u t}\left[y_0^u + \int_0^\infty e^{(-\mathcal{A}_N^u)\tau} P_N(\mathcal{A}PD)f(\tau)d\tau\right] + \rho(t) \\ \rho(t) = -\int_t^\infty e^{(-\mathcal{A}_N^u)(\tau-t)} P_N(\mathcal{A}PD)f(\tau)d\tau \end{cases} \tag{3.5.42}$$

$$= -\int_t^\infty e^{(-\mathcal{A}_N^u)(\tau-t)}(\mathcal{A}_N^u)P_N PDf(\tau)d\tau. \tag{3.5.43}$$

Indeed, setting $\mathcal{A}_N^u P_N PDf(t) = P_N(\mathcal{A}PD)f(t) \equiv h(t)$, we compute for the 2nd term in (3.5.38):

$$\int_0^t e^{\mathcal{A}_N^u(t-\tau)} h(\tau)d\tau = e^{\mathcal{A}_N^u t}\int_0^\infty e^{(-\mathcal{A}_N)\tau} h(\tau)d\tau - \int_t^\infty e^{(-\mathcal{A}_N^u)(\tau-t)} h(\tau)d\tau, \tag{3.5.44}$$

with $e^{(-\mathcal{A}_N^u)t}$ stable, and (3.5.44) used in the RHS of (3.5.38) yields (3.5.42), (3.5.43). Since $|e^{(-\mathcal{A}_N^u)t}| \leq C$, $t \geq 0$, and $f(t)$ decays as in (3.5.39), we obtain from (3.5.43),

$$|\rho(t)| \leq C\int_t^\infty e^{-a\tau}d\tau = \frac{C}{a}e^{-at}, \quad t \geq 0. \tag{3.5.45}$$

Then, (3.5.42), (3.5.43), (3.5.45) combined provide the desired conclusion. In fact, if (3.5.40) holds true, then (3.5.42) gives $r(t) = \rho(t)$, exponentially decaying by (3.5.45) and (3.5.41) is established. Conversely, if $r(t) \to 0$ as $t \to \infty$, then (3.5.42) implies (3.5.40) in view of (3.5.45) and the fact that $e^{\mathcal{A}_N^u t}$ is unstable; in which case, $r(t) = \rho(t)$, and the decay of $r(t)$ is actually exponential as given by (3.5.41) as $t \to \infty$, in view of (3.5.45). $\square$

3.5. THEOREM 2.1, GENERAL CASE $d = 3$

Step (v). We return to the $P\hat{y}$-dynamics (3.5.33) which is of the form (3.5.38) with: (i) $r(t) = P\hat{y}(t; y_0) - \eta(t; y_0)$, and (ii) $f(t) = -\tilde{y}(t; \tilde{y}_0)|_\Gamma$. We notice that $r(t)$ decays exponentially as $t \to \infty$ by invoking (3.5.24) on $\hat{y}(t; y_0)$ and (3.5.35) on $\eta(t; y_0)$, so that $a = (\gamma_0 - \epsilon)$ in (3.5.39) (since $0 < \gamma_0 < \tilde{\gamma}$), with $y_0 \in H^{\frac{1}{2}+\epsilon}(\Omega))^3 \cap H$, and so $\tilde{y}_0 \in \mathcal{D}(\tilde{\mathcal{A}}^{\frac{1}{4}+\frac{\epsilon}{2}})$. Similarly, $f(t) = -\tilde{y}(t; \tilde{y}_0)|_\Gamma$ decays exponentially by (3.5.21), and we can again take (conservatively) $a = \gamma_0 - \epsilon$ in (3.5.39). Moreover, $\int_\Gamma \tilde{y}(t; \tilde{y}_0) \cdot \nu = 0$ by Gauss theorem, as required, since $\nabla \cdot \tilde{y} \equiv 0$ in $\tilde{\Omega}$, see (3.5.10b). We can then apply the necessary condition of Lemma 3.5.2 to the $P\hat{y}$-dynamics (3.5.33). We thus obtain the present version of condition (3.5.40):

$$(3.5.46) \qquad y_0^u - \int_0^\infty e^{(-\mathcal{A}_N^u)\tau} P_N(\mathcal{APD})\tilde{y}(\tau; \tilde{y}_0)|_\Gamma d\tau = 0.$$

Step (vi). We now seek to apply the sufficient condition of Lemma 3.5.2 to the dynamics (3.5.36) for $y(t; y_0)$, with (i) $r(t) = y(t; y_0) - \zeta(t; y_0)$, and (ii) $f(t) = -g(t; \tilde{y}_0)$, which is exponentially stable in $(L^2(\Gamma))^3$ by (3.5.29), and satisfies the stronger pointwise boundary c.c. (3.5.28). Thus, (3.5.39) holds with $a = \tilde{\gamma}$ and $\tilde{y}_0 \in \mathcal{D}(\tilde{\mathcal{A}}^{\frac{1}{4}+\frac{\epsilon}{2}})$ in our case. Accordingly, we may conclude that $r(t) = y(t; y_0) - \zeta(t; y_0)$ decays exponentially as claimed by (3.5.41), that is

$$(3.5.47) \qquad |r(t)|_{(L^2(\Omega))^3} = |y(t; \tilde{y}_0) - \zeta(t; \tilde{y}_0)|_{(L^2(\Omega))^3} \leq Ce^{-\tilde{\gamma}t}|\tilde{y}_0|_{\mathcal{D}(\tilde{\mathcal{A}}^{\frac{1}{4}+\frac{\epsilon}{2}})}, \quad t \geq 0,$$

hence—in view of (3.5.37) for $\zeta(t; y_0)$,

$$(3.5.48) \qquad |y(t; \tilde{y}_0)|_H \leq Ce^{-(\gamma_0-\epsilon)t}|\tilde{y}_0|_{\mathcal{D}(\tilde{\mathcal{A}}^{\frac{1}{4}+\frac{\epsilon}{2}})}, \quad t \geq 0,$$

$(0 < \gamma_0 < \tilde{\gamma})$ as desired, *provided that* the present version of condition (3.5.40) holds true; that is, *provided that*

$$(3.5.49a) \qquad y_0^u - \int_0^\infty e^{(-\mathcal{A}_N^u)\tau} P_N(\mathcal{APD}) g(\tau; \tilde{y}_0) d\tau = 0,$$

or, recalling (3.5.27) for $g(\,\cdot\,)$, *provided that*

$$(3.5.49b) \qquad y_0^u - \int_0^\infty e^{(-\mathcal{A}_N^u)\tau} P_N(\mathcal{APD})[\tilde{y}(\tau; \tilde{y}_0)|_\Gamma - (\tilde{y}(\tau; \tilde{y}_0)|_\Gamma \cdot \nu)\nu] d\tau = 0.$$

But, in view of (3.5.46) already established, we see that the required condition (3.5.49b) holds true, *provided that* we can establish that

$$(3.5.50) \qquad P_N(\mathcal{APD})(\tilde{y}(t; \tilde{y}_0)|_\Gamma \cdot \nu)\nu = 0, \quad t \geq 0,$$

or, equivalently, that recalling $Z_N^u = P_N H$ from (3.4.4),

$$(3.5.51) \qquad ((P_N(\mathcal{APD})(\tilde{y}(t; \tilde{y}_0)|_\Gamma \cdot \nu)\nu, h^u)_{Z_N^u} = 0, \quad \forall\, h^u \in Z_N^u;$$

or—since $P_N^* h^u \in (Z_N^u)^* \equiv P_N^* H$, $\mathcal{A}^* P_N^* h^u \in (Z_N^u)^*$ so that $D^* P^* \mathcal{A}^* P_N^* h^u = D^* \mathcal{A}^* P_N^* h^u$, as the Leray orthogonal projector $P = P^*$ is redundant—equivalently that

$$(3.5.52) \qquad (\tilde{y}(t; \tilde{y}_0)|_\Gamma \cdot \nu)\nu,\, D^* \mathcal{A}^* P_N^* h^u)_{(L^2(\Gamma))^3} = 0, \quad \forall\, h^u \in Z_N^u;$$

48 3. PROOF OF THEOREMS 2.1 AND 2.2

or, since $P_N^* Z_N^u \subset \mathcal{D}(\mathcal{A}^*)$, and via identity (3.2.2), equivalently that

(3.5.53) $$\left((\tilde{y}(t;\tilde{y}_0)|_\Gamma \cdot \nu)\nu, \frac{\partial P_N^* h^u}{\partial \nu} \right)_{(L^2(\Gamma))^3} = 0;$$

$$\text{or } \int_\Gamma [(\tilde{y}(t;\tilde{y}_0)|_\Gamma \cdot \nu)\nu] \cdot \frac{\partial P_N^* h^u}{\partial \nu} \, d\Gamma = 0, \ \forall \ h^u \in Z_N^u.$$

Indeed, the required condition (3.5.53) does hold true. In fact, $P_N^* h^u \in (Z_N^u)^*$, where $(Z_N^u)^*$ is the span of all *generalized eigenfunctions* of $\mathcal{A}^*$ corresponding to its unstable distinct eigenvalues λ_j, $j = 1, \ldots, M$, see (3.6.3) below. Accordingly, we have: $P_N^* h^u|_\Gamma = 0$ and $\nabla \cdot P_N^* h^u \equiv 0$ in Ω. Thus, invoking Lemma 3.3.1, we conclude that

(3.5.54) $\quad \dfrac{\partial P_N^* h^u}{\partial \nu}$ is a vector tangential to Γ at every $x \in \Gamma$.

But, by definition, $(\tilde{y}(t;\tilde{y}_0)|_\Gamma \cdot \nu)\nu$ is a vector orthogonal to Γ at every $x \in \Gamma$. Thus, from here and (3.5.54), we conclude that the inner product integrand of the integral $\int_\Gamma \cdot \, d\Gamma$ appearing in (3.5.53) vanishes at each point of Γ. Accordingly, condition (3.5.53) is satisfied, as desired. We conclude that conditions (3.5.49), (3.5.50) hold true: $y(t;y_0)$ decays exponentially, as stated in (3.5.48). Thus, Proposition 3.5.2, Eqn. (3.5.31) is proved. □

Step 6. Next, we establish that for y-problem (3.5.26), (3.5.27), the required c.c. (recall (3.1.33)):

(3.5.55) $$y_0|_\Gamma = g(0;\tilde{y}_0)$$

holds true, for any $y_0 \in (H^{\frac{1}{2}+\epsilon}(\Omega))^3 \cap H$; that is, with no constraint on the initial condition. From (3.5.27), we compute,

(3.5.56) $\quad$ on $\Gamma : g(0;\tilde{y}_0) \ = \ \tilde{y}(0;\tilde{y}_0)|_\Gamma - (\tilde{y}(0;\tilde{y}_0)|_\Gamma \cdot \nu)\nu$

(3.5.57) $\qquad\qquad\qquad = \ \tilde{y}_0|_\Gamma - (\tilde{y}_0|_\Gamma \cdot \nu)\nu = y_0|_\Gamma - (y_0|_\Gamma \cdot \nu)\nu = y_0|_\Gamma,$

as desired. In going from (3.5.56) to (3.5.57) we have used the following properties: (i) that $\tilde{y}_0 \in (H_0^{\frac{1}{2}+\epsilon}(\tilde{\Omega}))^3 \cap \tilde{H}$ is an extension of $y_0 \in (H^{\frac{1}{2}+\epsilon}(\Omega))^3 \cap H$ from Ω to $\tilde{\Omega}$, constructed as in Step 1, so that $\tilde{y}_0|_{\Gamma=\partial\Omega} = y_0|_\Gamma$; (ii) that $y_0|_\Gamma \cdot \nu = 0$ on $\Gamma = \partial\Omega$, since $y_0 \in H \equiv H_\Omega$. Thus, (3.5.57) or (3.5.55) is proved.

Step 7. We can finally complete the proof of Theorem 3.5.1.

PROPOSITION 3.5.4. *Consider the y-problem (3.5.26), (3.5.27), with $y_0 \in (H^{\frac{1}{2}+\epsilon}(\Omega))^3 \cap H$, and ω being a strip all around the boundary $\Gamma = \partial\Omega$ of Ω, and $\tilde{\Omega} = \Omega \cup \omega$, as taken so far in Section 3.5, or, more generally, as specified in Remark 3.5.1, according to y_0. Then, we have, with Γ_1 as specified in Remark 3.5.1, depending on y_0:*

(3.5.58a) $\qquad g(t) \ \equiv \ \tilde{y}(t;\tilde{y}_0)|_\Gamma - (\tilde{y}(t;\tilde{y}_0|_\Gamma \cdot \nu)\nu, \ g \cdot \nu \equiv 0 \ on \ \Sigma,$

(3.5.58b) $\qquad\qquad \in \ H^{1+\epsilon, \frac{1}{2}+\frac{\epsilon}{2}}(\Sigma_{1,\infty})$

$\qquad\qquad \equiv \ L^2(0,\infty; (H^{1+\epsilon}(\Gamma_1))^3) \cap H^{\frac{1}{2}+\frac{\epsilon}{2}}(0,\infty; (L^2(\Gamma_1))^3)$

(3.5.58c) $\qquad g \equiv 0 \ on \ \Gamma \setminus \Gamma_1; \ g(0;\tilde{y}_0) = y_0|_\Gamma \in (H^\epsilon(\Gamma))^3,$

3.5. THEOREM 2.1, GENERAL CASE $d = 3$

and, continuously,

$$(3.5.59) \quad y \in H^{\frac{3}{2}+\epsilon, \frac{3}{4}+\frac{\epsilon}{2}}(Q_\infty) \equiv L^2\left(0, \infty; (H^{\frac{3}{2}+\epsilon}(\Omega))^3 \cap H\right) \cap H^{\frac{3}{4}+\frac{\epsilon}{2}}(0, \infty; H).$$

PROOF. The regularity of g, defined by (3.5.58a) (see (3.5.27)), was already established in (3.5.30), which then becomes (3.5.58b) with $g \equiv 0$ on $\Gamma \setminus \Gamma_1$, by recalling Remark 3.5.1. The c.c. in (3.5.58c) was established in (3.5.57). At this point, if we were only interested in the regularity of y only over a finite interval $[0, T]$, rather than the infinite interval $(0, \infty)$, we could just invoke Theorem 3.1.8 with $d = 3$ and $\theta = \frac{1}{2} + \frac{\epsilon}{2}$. In this case, our present g in (3.5.58b) fits (3.1.66); our present $y_0 \in (H^{\frac{1}{2}+\epsilon}(\Omega))^3 \cap H$ fits (3.1.68); our present c.c. (3.5.58c) fits the c.c. (3.1.69). Applying conclusion (3.1.70) of Theorem 3.1.8 with $\theta = \frac{1}{2} + \frac{\epsilon}{2}$, we would then get (3.5.59), except on any finite interval $(0, T)$.

To get the desired conclusion (3.5.59) on an infinite time interval, we proceed as follows.

(i) First, by considering the stable component $y^s = (I - P_N)y$ of y, and applying the proof of Theorem 3.8.1 in this stable case, we obtain

$$(3.5.60) \quad (I - P_N)y = y^s \in L^2\left(0, \infty; (H^{\frac{3}{2}+\epsilon}(\Omega))^3 \cap H\right) \cap H^{\frac{3}{4}+\frac{\epsilon}{2}}(0, \infty; H).$$

(ii) It remains to show that, with $\tilde{y}_0 \in (H_0^{\frac{1}{2}+\epsilon}(\tilde{\Omega}))^3 \cap \tilde{H} = \mathcal{D}(\tilde{A}^{\frac{1}{4}+\epsilon})$, see (1.15), we have

$$(3.5.61) \quad P_N y = y^u \in L^2\left(0, \infty; (H^{\frac{3}{2}+\epsilon}(\Omega))^3 \cap H\right) \cap H^{\frac{3}{4}+\frac{\epsilon}{2}}(0, \infty; H).$$

But from Proposition 3.5.2, Eqn. (3.5.31), we see that with such Initial Condition, we have

$$(3.5.62) \quad y^u \in L^2(0, \infty; Z_N^u), \quad Z_N^u = P_N Z \text{ with the topology of } H,$$

and Z_N^u is finite-dimensional. Thus on Z_N^u, the H-topology and the $(H^{\frac{3}{2}+\epsilon}(\Omega))^3$-topology are equivalent. Hence, the space regularity in (3.5.61) is established, and similarly for the time regularity, since y^u is the solution of a linear, finite-dimensional system on the finite-dimensional space $P_N Z \equiv Z_N^u$. Hence, $y^u(t) = e^{Ft} y_0^u$, for a suitable matrix F. □

Step (iii). Combining (3.5.60) for the stable component y^s and (3.5.61) for the unstable component y^u, we finally obtain (3.5.59) for $y = y^s + y^u$. The proof of Proposition 3.5.4 is complete. □

In conclusion, the proof of Theorem 3.5.1 is complete. □

REMARK 3.5.3. To appreciate Theorem 3.1.8 (and, thus, Theorem 3.1.4 and Theorem 3.1.7 that lead to it), we notice what follows:

(1) A direct use of formula (3.5.32b) for y will not do: at this high topological level, where $\mathcal{D}((\mu - \mathcal{A})^{\frac{3}{4}+\frac{\epsilon}{2}})$ includes the zero Dirichlet B.C., $\mathcal{D}((\mu - \mathcal{A})^{\frac{3}{4}+\frac{\epsilon}{2}}) = (H^{\frac{3}{2}+\epsilon}(\Omega))^3 \cap V$, see (1.17), the integral term in formula (3.5.32b) is not suitable for distributing fractional powers of $(\mu - \mathcal{A})$ across. More precisely, we start from the regularity $\tilde{y}(t; \tilde{y}_0) \in L^2(0, \infty; (H^{\frac{3}{2}+\epsilon}(\tilde{\Omega}))^3 \cap \tilde{V})$ in (3.5.16), apply trace theory to

get $\tilde{y}(t; y_0)|_\Gamma$ hence $g \in L^2(0, \infty; (H^{1+\epsilon}(\Gamma))^3)$ by (3.5.27), and finally invoke (3.1.3d) for D to obtain, via (3.5.27) on g that

(3.5.63a) $\quad Dg(t; \tilde{y}_0) \in L^2(0, \infty; (H^{\frac{3}{2}+\epsilon}(\Omega))^3 \cap H)$, but

(3.5.63b) $\quad Dg(t; \tilde{y}_0) \notin L^2(0, \infty; \mathcal{D}(\mu I - \mathcal{A})^{\frac{3}{4}+\frac{\epsilon}{2}}) = L^2(0, \infty; (H^{\frac{3}{2}+\epsilon}(\Omega))^3 \cap V)$,

by (1.17), since $Dg|_\Gamma = 0$ if and only if $g = 0$ on Γ by its definition. Thus, this approach fails as the topology is *too high*.

(2) If, instead, one attempts to use directly the solution formula (3.1.13) [or (3.1.38)] with $\tilde{y}_t(t; \tilde{y}_0) \in L^2(0, \infty; (H^{-\frac{1}{2}+\epsilon}(\tilde{\Omega}))^3)$, see (3.5.17), we are at a *too low* topological level to take the trace of $\tilde{y}_t$ on $\Gamma \equiv \partial\Omega$. Thus, this approach also fails.

It is the balance of g in *time and space* as described in (3.5.30) or (3.5.58a) that succeeds via (Theorem 3.1.4, Theorem 3.1.7, hence) Theorem 3.1.8 to provide the sought-after regularity (3.5.59) for y (even on a finite time interval).

REMARK 3.5.4. Thus, this way, by virtue of Theorem 3.5.1, we have managed to satisfy (in an open-loop form) the Finite Cost Condition of the optimal control problem in Section 4:

(3.5.64) $\quad \int_0^\infty \left[|y(t; y_0)|^2_{\frac{3}{2}+2\epsilon} + |g(t)|^2 \right] dt < \infty, \; y \cdot \nu \equiv 0 \text{ on } \Sigma$

$$\text{for any } y_0 \in \left(H^{\frac{1}{2}+2\epsilon}(\Omega) \right)^3 \cap H,$$

for $d = 3$, in full generality. This way, one may begin the analysis of the corresponding Riccati theory for such optimal control problem, to be carried out in Section 4.

3.6. Feedback stabilization of the unstable z_N-system (3.4.9) on Z_N^u under the FDSA

Orientation. Beginning with the present Section 3.6 and concluding with Section 3.7, we carry out an alternative approach to satisfy the FCC (3.1.22)–(3.1.24) of the linearized model (3.1.2), or (3.1.14a), with $d = 2, 3$. As explained in the Orientation of Section 3.5, this alternative approach produces a required open-loop boundary control with additional desirable properties: (i) the boundary control g has an arbitrarily small support on Γ; (ii) the difference $[g(t) - e^{-2\gamma_1 t} y_0|_\Gamma]$ is finite-dimensional. These extra features of the sought-after boundary control come, however, at the price of making a finite-dimensional spectral assumption, FDSA, to be spelled out below in (3.6.2). We follow [**T.1**], [**T.2**].

We now denote by $\{\varphi_{ij}\}_{j=1}^{\ell_i}$, $\{\varphi_{ij}^*\}_{j=1}^{\ell_i}$, the (normalized) linearly independent eigenfunctions corresponding to each unstable distinct eigenvalue λ_i of $\mathcal{A}$ and $\bar{\lambda}_i$ of $\mathcal{A}^*$, respectively:

(3.6.1) $\quad\quad\quad \mathcal{A}\varphi_{ij} = \lambda_i \varphi_{ij}; \quad \mathcal{A}^* \varphi_{ij}^* = \bar{\lambda}_i \varphi_{ij}^*.$

We now introduce the *a finite-dimensional spectral assumption*:

(FDSA): We assume that for each of the distinct unstable eigenvalues $\lambda_1, \ldots, \lambda_M$ of $\mathcal{A}$, algebraic and geometric multiplicity coincide:

(3.6.2) $\quad Z_{N,i} \equiv P_{N,i} H = \text{span}\{\varphi_{ij}\}_{j=1}^{\ell_i};$

$\quad\quad\quad (Z_{N,i})^* \equiv P_{N,i}^* H = \text{span}\{\varphi_{ij}^*\}_{j=1}^{\ell_i}; \; \ell_1 + \ell_2 + \cdots + \ell_M = N,$

where $P_{N,i}$, $P_{N,i}^*$ are the projection defined as in (3.4.3) after replacing Γ, $\bar{\Gamma}$ with closed curves Γ_i, $\bar{\Gamma}_i$ enclosing λ_i and $\bar{\lambda}_i$, and no other point of the spectrum of $\mathcal{A}$ and $\mathcal{A}^*$, respectively. As a consequence of the FDSA, we obtain

(3.6.3) $Z_N^u = P_N H = \text{span}\{\varphi_{ij}\}_{i=1}^M {}_{j=1}^{\ell_i}$; $(Z_N^u)^* = P_N^* H = \text{span}\{\varphi_{ij}^*\}_{i=1}^M {}_{j=1}^{\ell_i}$.

[without the FDSA, Z_N^u is the span of the *generalized* eigenfunctions of $\mathcal{A}$ corresponding to its unstable eigenvalues; and similarly for $(Z_N^u)^*$. The latter property was used above (3.5.54).]

In other words, the FDSA says that the restriction $\mathcal{A}_N^u$ in (3.4.5) is *diagonalizable* or that $\mathcal{A}_N^u$ is a normal operator on Z_N^u. In the terminology of [**K.1**], $\mathcal{A}_N^u$ is *semi-simple*. Another equivalent characterization of the FDSA is that: each unstable eigenvalue $\lambda_1, \ldots, \lambda_N$ is a simple pole (pole of order 1) of the resolvent $R(\lambda, \mathcal{A})$ [**K.1**, p. 43].

REMARK 3.6.1. It appears that no simple sufficient conditions are known that guarantee the validity of the FDSA. A general feeling is that the FDSA may be 'generic'; the same way that, generically, the Laplacian Δ on an Euclidean bounded domain Ω (or Riemannian manifold) with classical boundary conditions has simple eigenvalues [**A.1**], [**M.1**], [**U.1**].

REMARK 3.6.2. In our space-decomposition development, the FDSA is used critically only in one spot: in claiming that for each $i = 1, \ldots, M$, the normal traces $\{\frac{\partial}{\partial \nu} \varphi_{ij}\}_{j=1}^{\ell_i}$ of the eigenfunctions φ_{ij} of $\mathcal{A}$ corresponding to its unstable distinct eigenvalues λ_i are linearly independent in $(L^2(\Gamma))^d$; in fact, even on $(L^2(\Gamma_1))^d$ for an arbitrary subportion Γ_1 of Γ of positive (surface) measure; see Eqn. (3.6.20) below.

If we, instead, let $\{\varphi_{ij}\}_{j=1}^{\ell_i}$ be the *generalized* eigenfunctions of $\mathcal{A}$, corresponding to its unstable eigenvalue λ_i with *algebraic* multiplicity ℓ_i—hence $Z_{N,j} = \{z \in H; \ (\lambda_j - \mathcal{A})^{\ell_j} z = 0\}$, $j = 1, \ldots, M$, instead of (3.6.1)—we are unable to claim that the corresponding normal traces $\{\frac{\partial}{\partial \nu} \varphi_{ij}\}_{j=1}^{\ell_i}$ are linearly independent in $(L^2(\Gamma))^d$. This is so since a uniqueness result from [**B-T.1**, Lemma 3.7]— see also [**F.2**], [**F.3**]—which is invoked below to prove (3.6.20) and which holds true for a (true) eigenfunction φ of $\mathcal{A}$ with over-determined boundary conditions: $\varphi|_\Gamma = \frac{\partial \varphi}{\partial \nu}|_\Gamma = 0 \Rightarrow \varphi \equiv 0$, appears to be false in the case of a generalized eigenfunction φ, see Remark 3.6.4.

Implications of the FDSA = (3.6.2). In this section, we assume henceforth the FDSA = (3.6.2). Then, we have for $i = 1, \ldots, M$, $j = 1, \ldots, \ell_i$:

(3.6.4) $\mathcal{A}_N^u \varphi_{ij} = \lambda_i \varphi_{ij}$; $(\mathcal{A}_N^u)^* \varphi_{ij}^* = \bar{\lambda}_i \varphi_{ij}^*$; $\ell_1 + \ell_2 + \cdots + \ell_M = N$.

In this case, $\mathcal{A}_N^u$ is diagonal with respect to the basis $\{\varphi_{ij}\}$. It follows that the system consisting of $\{\varphi_{ij}\}$ and $\{\varphi_{ij}^*\}$, $i = 1, \ldots, M$, can be chosen to form biorthogonal sequences: they can be selected in such a way that the following biorthogonal relations hold true in the H-inner product [**B-T.1**, Eqn. (3.7)], [**K.1**, p. 51]

(3.6.5) $(\varphi_{ij}, \varphi_{hk}^*)_H = \begin{cases} 0 & \text{if } i \neq h; \text{ or if } i = h, j \neq k; \\ 1 & \text{if } i = h, j = k, \end{cases}$

i and h running from 1 through M, and j and k running from 1 to ℓ_i and ℓ_k, respectively. It then follows from (3.6.5) that any vector $z \in Z_N^u$ admits the unique expansion [**K.1**, p. 12]

(3.6.6) $\quad Z_N^u \ni z = \sum_{i,j}(z, \varphi_{ij}^*)_H \varphi_{ij}, \quad i = 1, \ldots, M, \; j = 1, \ldots, \ell_i.$

In the present case, we then have the following explicit representation of $P_N(\mathcal{A}Dv) = \mathcal{A}_N^u P_N Dv$, needed in (3.4.9). First, let $v \in (L^2(\Gamma))^d$, $v \cdot \nu \equiv 0$ on Γ, so that (3.1.3d) holds true for $s = 0$. Next, see (3.4.3) or (3.6.2), since $P_N^* : H$ onto $(Z_N^u)^* = \text{span}\{\varphi_{ij}^*\}_{i=1}^M{}_{j=1}^{\ell_i}$, we have $P_N^* \varphi_{ij}^* = \varphi_{ij}^* \in \mathcal{D}(\mathcal{A}^*)$. Since the fundamental identity (3.2.2) holds true due to our standing assumption (3.1.2e) $v \cdot \nu \equiv 0$ on Σ, we have via (3.6.6) in the duality pairing:

$$\text{(3.6.7)} \quad P_N(\mathcal{A}Dv) = \sum_{i,j=1}^{M,\ell_i}(P_N(\mathcal{A}Dv), \varphi_{ij}^*)_H \varphi_{ij}$$

$$= \sum_{i,j}(\mathcal{A}Dv, \varphi_{ij}^*)_H \varphi_{ij}$$

$$\text{(by (3.2.2))} \quad = \sum_{i,j}(v, D^*\mathcal{A}^*\varphi_{ij}^*)_{(L_2^2(\Gamma))^d} \varphi_{ij}$$

$$\text{(3.6.8)} \quad = \nu_0 \sum_{i,j=1}^{M,\ell_i}\left(v, \left.\frac{\partial \varphi_{ij}^*}{\partial \nu}\right|_\Gamma\right)_{(L^2(\Gamma))^d} \varphi_{ij}.$$

Motivated by (3.6.8), we introduce the following subspace:

(3.6.9) $\quad \mathcal{F} \equiv \text{span}\left\{\frac{\partial}{\partial \nu}\varphi_{ij}^*, \; i = 1, \ldots, M; \; j = 1, \ldots, \ell_i\right\} \subset (H^{\frac{1}{2}}(\Gamma))^d,$

where containment follows by trace theory on $\varphi_{ij}^* \in (H^2(\Omega))^d$. Moreover, if $w_1, w_2, \ldots, w_k \in (L^2(\Gamma))^d$, we introduce the $\ell_i \times k$ matrix W_i, for $i = 1, \ldots, M$:

$$\text{(3.6.10)} \quad W_i = \begin{bmatrix} (w_1, \partial_\nu \varphi_{i1}^*)_\Gamma & \cdots, & (w_k, \partial_\nu \varphi_{i1}^*)_\Gamma \\ (w_1, \partial_\nu \varphi_{i2}^*)_\Gamma & \cdots, & (w_k, \partial_\nu \varphi_{i2}^*)_\Gamma \\ \vdots & & \vdots \\ (w_1, \partial_\nu \varphi_{i\ell_i}^*)_\Gamma & \cdots, & (w_k, \partial_\nu \varphi_{i\ell_i}^*)_\Gamma \end{bmatrix},$$

$$\partial_\nu = \frac{\partial}{\partial \nu}, \quad (\;,\;)_\Gamma = (\;,\;)_{(L^2(\Gamma))^d}.$$

LEMMA 3.6.1. *Assume the FDSA = (3.6.2). Given $\gamma_1 > 0$ arbitrarily large, there is a controller $v = v_N = \sum_{i=1}^K v_N^i(t) w_i$, $K \leq N$ (conservatively) for infinitely many suitable vectors $w_i \in \mathcal{F}$, see (3.6.9), $w_i \cdot \nu \equiv 0$ on Γ, which will have to satisfy the rank conditions (3.6.16) below, such that once inserted in (3.4.9), yields the estimate*

(3.6.11) $\quad |z_N(t)| + |v_N(t)|_{(H^{\frac{1}{2}}(\Gamma))^d} + |\dot{v}_N(t)|_{(H^{\frac{1}{2}}(\Gamma))^d} \leq C_{\gamma_1} e^{-\gamma_1 t} |P_N z_0|, \; t \geq 0.$

Moreover, the vectors w_i can be chosen to have support on an arbitrarily small portion Γ_1 of Γ. Here z_N is the solution to (3.4.9) corresponding to such v. Moreover, the controller $v = v_N$ may be chosen in feedback form: that is, of the form where $P_{N,i}$, $P_{N,i}^*$ are the projection defined as in (3.4.3) after replacing Γ, $\bar{\Gamma}$ with closed curves Γ_i, $\bar{\Gamma}_i$ enclosing λ_i and $\bar{\lambda}_i$, and no other point of the spectrum of $\mathcal{A}$ and $\mathcal{A}^*$, respectively. As a consequence of the FDSA, we obtain

(3.6.3) $\quad Z_N^u = P_N H = \text{span}\{\varphi_{ij}\}_{i=1}^{M}{}_{j=1}^{\ell_i}; \quad (Z_N^u)^* = P_N^* H = \text{span}\{\varphi_{ij}^*\}_{i=1}^{M}{}_{j=1}^{\ell_i}.$

[without the FDSA, Z_N^u is the span of the *generalized* eigenfunctions of $\mathcal{A}$ corresponding to its unstable eigenvalues; and similarly for $(Z_N^u)^*$. The latter property was used above (3.5.54).]

In other words, the FDSA says that the restriction $\mathcal{A}_N^u$ in (3.4.5) is *diagonalizable* or that $\mathcal{A}_N^u$ is a normal operator on Z_N^u. In the terminology of [**K.1**], $\mathcal{A}_N^u$ is *semi-simple*. Another equivalent characterization of the FDSA is that: each unstable eigenvalue $\lambda_1, \ldots, \lambda_N$ is a simple pole (pole of order 1) of the resolvent $R(\lambda, \mathcal{A})$ [**K.1**, p. 43].

REMARK 3.6.1. It appears that no simple sufficient conditions are known that guarantee the validity of the FDSA. A general feeling is that the FDSA may be 'generic'; the same way that, generically, the Laplacian Δ on an Euclidean bounded domain Ω (or Riemannian manifold) with classical boundary conditions has simple eigenvalues [**A.1**], [**M.1**], [**U.1**].

REMARK 3.6.2. In our space-decomposition development, the FDSA is used critically only in one spot: in claiming that for each $i = 1, \ldots, M$, the normal traces $\{\frac{\partial}{\partial \nu} \varphi_{ij}\}_{j=1}^{\ell_i}$ of the eigenfunctions φ_{ij} of $\mathcal{A}$ corresponding to its unstable distinct eigenvalues λ_i are linearly independent in $(L^2(\Gamma))^d$; in fact, even on $(L^2(\Gamma_1))^d$ for an arbitrary subportion Γ_1 of Γ of positive (surface) measure; see Eqn. (3.6.20) below.

If we, instead, let $\{\varphi_{ij}\}_{j=1}^{\ell_i}$ be the *generalized* eigenfunctions of $\mathcal{A}$, corresponding to its unstable eigenvalue λ_i with *algebraic* multiplicity ℓ_i—hence $Z_{N,j} = \{z \in H; (\lambda_j - \mathcal{A})^{\ell_j} z = 0\}$, $j = 1, \ldots, M$, instead of (3.6.1)—we are unable to claim that the corresponding normal traces $\{\frac{\partial}{\partial \nu} \varphi_{ij}\}_{j=1}^{\ell_i}$ are linearly independent in $(L^2(\Gamma))^d$. This is so since a uniqueness result from [**B-T.1**, Lemma 3.7]—see also [**F.2**], [**F.3**]—which is invoked below to prove (3.6.20) and which holds true for a (true) eigenfunction φ of $\mathcal{A}$ with over-determined boundary conditions: $\varphi|_\Gamma = \frac{\partial \varphi}{\partial \nu}|_\Gamma = 0 \Rightarrow \varphi \equiv 0$, appears to be false in the case of a generalized eigenfunction φ, see Remark 3.6.4.

Implications of the FDSA = (3.6.2). In this section, we assume henceforth the FDSA = (3.6.2). Then, we have for $i = 1, \ldots, M$, $j = 1, \ldots, \ell_i$:

(3.6.4) $\quad \mathcal{A}_N^u \varphi_{ij} = \lambda_i \varphi_{ij}; \quad (\mathcal{A}_N^u)^* \varphi_{ij}^* = \bar{\lambda}_i \varphi_{ij}^*; \quad \ell_1 + \ell_2 + \cdots + \ell_M = N.$

In this case, $\mathcal{A}_N^u$ is diagonal with respect to the basis $\{\varphi_{ij}\}$. It follows that the system consisting of $\{\varphi_{ij}\}$ and $\{\varphi_{ij}^*\}$, $i = 1, \ldots, M$, can be chosen to form biorthogonal sequences: they can be selected in such a way that the following biorthogonal relations hold true in the H-inner product [**B-T.1**, Eqn. (3.7)], [**K.1**, p. 51]

(3.6.5) $\quad (\varphi_{ij}, \varphi_{hk}^*)_H = \begin{cases} 0 & \text{if } i \neq h; \text{ or if } i = h, j \neq k; \\ 1 & \text{if } i = h, j = k, \end{cases}$

i and h running from 1 through M, and j and k running from 1 to ℓ_i and ℓ_k, respectively. It then follows from (3.6.5) that any vector $z \in Z_N^u$ admits the unique expansion [**K.1**, p. 12]

(3.6.6) $$Z_N^u \ni z = \sum_{i,j}(z,\varphi_{ij}^*)_H \varphi_{ij}, \quad i=1,\ldots,M,\ j=1,\ldots,\ell_i.$$

In the present case, we then have the following explicit representation of $P_N(\mathcal{A}Dv) = \mathcal{A}_N^u P_N Dv$, needed in (3.4.9). First, let $v \in (L^2(\Gamma))^d$, $v \cdot \nu \equiv 0$ on Γ, so that (3.1.3d) holds true for $s=0$. Next, see (3.4.3) or (3.6.2), since $P_N^* : H$ onto $(Z_N^u)^* = \mathrm{span}\{\varphi_{ij}^*\}_{i=1\ j=1}^{M\ \ \ell_i}$, we have $P_N^* \varphi_{ij}^* = \varphi_{ij}^* \in \mathcal{D}(\mathcal{A}^*)$. Since the fundamental identity (3.2.2) holds true due to our standing assumption (3.1.2e) $v \cdot \nu \equiv 0$ on Σ, we have via (3.6.6) in the duality pairing:

(3.6.7) $$P_N(\mathcal{A}Dv) = \sum_{i,j=1}^{M,\ell_i} (P_N(\mathcal{A}Dv), \varphi_{ij}^*)_H \varphi_{ij}$$

$$= \sum_{i,j}(\mathcal{A}Dv, \varphi_{ij}^*)_H \varphi_{ij}$$

(by (3.2.2)) $$= \sum_{i,j}(v, D^*\mathcal{A}^*\varphi_{ij}^*)_{(L_2^2(\Gamma))^d} \varphi_{ij}$$

(3.6.8) $$= \nu_0 \sum_{i,j=1}^{M,\ell_i} \left(v, \left.\frac{\partial \varphi_{ij}^*}{\partial \nu}\right|_\Gamma\right)_{(L^2(\Gamma))^d} \varphi_{ij}.$$

Motivated by (3.6.8), we introduce the following subspace:

(3.6.9) $$\mathcal{F} \equiv \mathrm{span}\left\{\frac{\partial}{\partial \nu}\varphi_{ij}^*,\ i=1,\ldots,M;\ j=1,\ldots,\ell_i\right\} \subset (H^{\frac{1}{2}}(\Gamma))^d,$$

where containment follows by trace theory on $\varphi_{ij}^* \in (H^2(\Omega))^d$. Moreover, if $w_1, w_2, \ldots, w_k \in (L^2(\Gamma))^d$, we introduce the $\ell_i \times k$ matrix W_i, for $i=1,\ldots,M$:

(3.6.10) $$W_i = \begin{bmatrix} (w_1, \partial_\nu \varphi_{i1}^*)_\Gamma & \cdots & (w_k, \partial_\nu \varphi_{i1}^*)_\Gamma \\ (w_1, \partial_\nu \varphi_{i2}^*)_\Gamma & \cdots & (w_k, \partial_\nu \varphi_{i2}^*)_\Gamma \\ \vdots & & \vdots \\ (w_1, \partial_\nu \varphi_{i\ell_i}^*)_\Gamma & \cdots & (w_k, \partial_\nu \varphi_{i\ell_i}^*)_\Gamma \end{bmatrix},$$

$$\partial_\nu = \frac{\partial}{\partial \nu}, \quad (\ ,\ )_\Gamma = (\ ,\ )_{(L^2(\Gamma))^d}.$$

LEMMA 3.6.1. *Assume the FDSA = (3.6.2). Given $\gamma_1 > 0$ arbitrarily large, there is a controller $v = v_N = \sum_{i=1}^K v_N^i(t) w_i$, $K \leq N$ (conservatively) for infinitely many suitable vectors $w_i \in \mathcal{F}$, see (3.6.9), $w_i \cdot \nu \equiv 0$ on Γ, which will have to satisfy the rank conditions (3.6.16) below, such that once inserted in (3.4.9), yields the estimate*

(3.6.11) $$|z_N(t)| + |v_N(t)|_{(H^{\frac{1}{2}}(\Gamma))^d} + |\dot{v}_N(t)|_{(H^{\frac{1}{2}}(\Gamma))^d} \leq C_{\gamma_1} e^{-\gamma_1 t} |P_N z_0|,\ t \geq 0.$$

Moreover, the vectors w_i can be chosen to have support on an arbitrarily small portion Γ_1 of Γ. Here z_N is the solution to (3.4.9) corresponding to such v. Moreover, the controller $v = v_N$ may be chosen in feedback form: that is, of the form

3.6. FEEDBACK STABILIZATION OF THE UNSTABLE z_N-SYSTEM

$v_N^i(t) = (z_N(t), p_i)_H$, with suitable vectors $p_i \in Z_N^u$, depending on γ_1. In conclusion, z_N in (3.6.11) is the solution of the equation on Z_N^u

$$(3.6.12) \qquad z_N' - \mathcal{A}_N^u z_N = -\mathcal{A}_N^u P_N D \left(\sum_{i=1}^K (z_N(t), p_i)_H w_i \right)$$

rewritten as: $z_N' = \bar{A}^u z_N$, $z_N(0) = P_N z_0$, $z_N(t) = e^{\bar{A}^u t} P_N z_0$.

PROOF. Let at first $w_1, \ldots, w_K \in (L^2(\Gamma))^d$, $w_i \cdot \nu = 0$ on Γ and $v = \sum_{k=1}^K v_k(t) w_k$. Under the hypothesis FDSA = (3.6.2), we invoke (3.6.8) and obtain

$$(3.6.13) \qquad P_N(\mathcal{A}Dv) = P_N \mathcal{A} D \left(\sum_{k=1}^K v_k(t) w_k \right)$$

$$= \nu_0 \sum_{i,j=1}^{M, \ell_i} \left\{ \sum_{k=1}^K (w_k, \partial_\nu \varphi_{ij}^*)_\Gamma v_k(t) \right\} \varphi_{ij},$$

which substituted into the RHS of (3.4.9) yields

$$(3.6.14\text{a}) \qquad z_N' - \mathcal{A}_N^u z_N = -\nu_0 \sum_{i=1}^M \sum_{j=1}^{\ell_i} \left\{ \sum_{k=1}^K (w_k, \partial_\nu \varphi_{ij}^*)_\Gamma v_k \right\} \varphi_{ij},$$

Next, we represent z_N as $z_N = \sum_{j=1}^N z_N^j \varphi_j = \sum_{ij} z_N^{ij} \varphi_{ij}$, $i = 1, \ldots, M$, $j = 1, \ldots, \ell_i$, and we set $\tilde{z}_N = \text{col}[z_N^1, \ldots, z_N^N]$. Then, in $\mathbb{C}^N$, with respect to the basis $\{\varphi_j\}_{j=1}^N$ of normalized eigenfunctions of $\mathcal{A}_N^u$, we may rewrite system (3.6.14a) by virtue of (3.6.4) for $\mathcal{A}$ as

$$(3.6.14\text{b}) \qquad (\tilde{z}_N)' - \Lambda \tilde{z}_N = W v_N,$$

$$(3.6.15) \qquad W = \|(w_r, \partial_\nu \varphi_{ij}^*)_\Gamma\| = \begin{bmatrix} W_1 \\ W_2 \\ \vdots \\ W_M \end{bmatrix}, \quad r = 1, \ldots, K, \text{ and}$$

$$\Lambda = \begin{bmatrix} \lambda_1 I_1 & & & 0 \\ & \lambda_2 I_2 & & \\ & & \ddots & \\ 0 & & & \lambda_M I_M \end{bmatrix},$$

$W : N \times K$; $\Lambda : N \times N$; $W_i = \ell_i \times K$. Then we have the following feedback stabilization result for the finite dimensional, unstable projected system (3.4.9) or (3.6.14a). $\square$

LEMMA 3.6.2. *Assume the FDSA = (3.6.2) and, moreover, that $k = K = N$ and that*

$$(3.6.16) \qquad \text{rank } W_i = \ell_i, \quad i = 1, \ldots, M,$$

where W_i is given by (3.6.10). Given $\gamma_1 > 0$ arbitrarily large, there are vectors $\{p_1, p_2, \ldots, p_K\} \in Z_N^u$, such that Eqn. (3.6.14a) with $v_k = (z_N, p_k)_H$, that is,

$$(3.6.17) \qquad z_N' - \mathcal{A}_N^u z_N = -\nu_0 \sum_{i=1, j=1}^{M, \ell_i} \left\{ \sum_{k=1}^K (w_k, \partial_\nu \varphi_{ij}^*)_\Gamma (z_N, p_k)_H \right\} \varphi_{ij}$$

rewritten as

$$(3.6.18) \qquad z_N' \equiv \bar{A}^u z_N, \quad \text{or} \quad z_N(t) = e^{\bar{A}^u t} z_N(0) = e^{\bar{A}^u t} P_N z_0$$

is exponentially stable with arbitrary decay rate $e^{-\gamma_1 t}$ in the precise sense stated by (3.6.11) [actually, $\bar{A}^u$ has an arbitrarily preassigned spectrum].

PROOF OF LEMMA 3.6.2. We shall now test the controllability of the pair $\{\Lambda, W\}$ in system (3.6.14b). To this end, we can either invoke the well-known Kalman controllability criterion on a generalized Vandermonde determinant, or else the Hautus controllability criterion: rank $[\Lambda - \lambda_i I, W] = N$ (full), for all eigenvalues λ_i of Λ. The latter criterion is actually simpler than the former, though both work successfully (see [**B-T.1**]). A direct application of Hautus criterion readily yields that the pair $\{\Lambda, W\}$ is controllable if and only if rank $W_i = \ell_i$ (full), $i = 1, \ldots, M$, which holds true by assumption (3.6.16). Hence, by the well-known Popov's criterion in finite-dimensional theory, there exists a (feedback) matrix $Q : K \times N$, such that the spectrum of the matrix $[\Lambda + WQ]$ may be arbitrarily preassigned; in particular, to lie in the left half-plane $\{\lambda : \operatorname{Re} \lambda < -\gamma_1 < -\operatorname{Re} \lambda_{N+1}\}$, as desired. The resulting closed loop system: $(\tilde{z}_N)' - \Lambda \tilde{z}_N = W v_N$ is obtained with $v_N = Q \tilde{z}_N$, Q being a $K \times N$ matrix with row vectors $[\tilde{p}_1, \ldots, \tilde{p}_K]$, whereby then $v_N^k = (\tilde{z}_N, \tilde{p}_k)$ in the $\mathbb{C}^N$-inner product. Thus, returning from $\mathbb{C}^N$ back to Z_N^u, there exist suitable vectors $p_1, \ldots, p_K$ in Z_N^u, such that $v_N^k = (z_N, p_k)$, whereby the closed loop system corresponding to (3.6.14a) is given precisely by (3.6.17), as claimed. In conclusion: under assumption (3.6.16), *a-fortiori*, given any $\gamma_1 > 0$, there exist vectors $\{p_1, \ldots, p_K\} \subset Z_N^u$ such that the solution z_N to (3.6.17) satisfies the estimate

$$(3.6.19) \quad |v_N(t)| + |z_N(t)| \;=\; |Q z_N(t)| + |z_N(t)|$$
$$\leq \;(|Q| + 1) |e^{\bar{A}^u t} z_N(0)| \leq C_{\gamma_1} e^{-\gamma_1 t} |z_N(0)|.$$

$\square$

COMPLETION OF PROOF OF LEMMA 3.6.1. To complete the proof of Lemma 3.6.1, we need to check that condition (3.6.16) is always possible in the present situation. In fact, we shall show that it can always be satisfied by infinitely many choices of the vectors $w_1, \ldots, w_K$ where conservatively we may take $K = N$.

Step 1. To this end, via (3.6.10), our key goal is the following: for each $i = 1, \ldots, M$,

(3.6.20a) \qquad the system $\{\partial_\nu(\varphi_{i,j}^*)\}_{j=1}^{\ell_i}$ is linearly independent in $(L^2(\Gamma))^d$.

In fact, even more is true: let Γ_1 be any portion of $\Gamma = \partial \Omega$ of positive surface measure. Then

(3.6.20b) \qquad the system $\{\partial_\nu(\varphi_{i,j}^*)\}_{j=1}^{\ell_i}$ is linearly independent in $(L^2(\Gamma_1))^d$.

$\square$

PROOF OF (3.6.20A). Under the FDSA = (3.6.2), the $\{\varphi_{ij}^*\}_{j=1}^{\ell_i}$ are true eigenfunctions corresponding to the eigenvalue $\bar{\lambda}_i$ of $\mathcal{A}^*$ as in (3.6.1), (3.6.4): $(\mathcal{A}_N^u)^*\varphi_{ij}^* = \mathcal{A}^*\varphi_{ij}^* = \bar{\lambda}_i \varphi_{ij}^*$, $i = 1, \ldots, M$; $j = 1, \ldots, \ell_i$. In this case, the validity of (3.6.20a) follows from the unique continuation result [**B-T.1**, Lemma 3.7], see also [**F.2**], [**F.3**]. In fact, with i arbitrary but fixed, let

$$(3.6.21) \qquad \sum_{j=1}^{\ell_i} \alpha_j \partial_\nu(\varphi_{ij}^*) = \partial_\nu \left(\sum_{j=1}^{\ell_i} \alpha_j \varphi_{ij}^* \right) = \partial_\nu \varphi^* = 0,$$

and show $\alpha_j \equiv 0$, $j = 1, \ldots, \ell_i$,

where we have defined $\varphi^* = \sum_{j=1}^{\ell_i} \alpha_j \varphi_{ij}^* \in (Z_{N,i})^*$, see (3.6.2). By the present FDSA = (3.6.2) assumption, we have that φ^* is a true eigenfunction of $\mathcal{A}^*$ corresponding to the eigenvalue $\bar{\lambda}_i$. Thus we have:

$$(3.6.22) \quad \mathcal{A}^*\varphi^* = \bar{\lambda}_i \varphi^*, \quad \varphi^*|_\Gamma = 0, \quad \left.\frac{\partial \varphi^*}{\partial \nu}\right|_\Gamma = 0, \ \varphi^* \in \mathcal{D}(\mathcal{A}^*) = (H^2(\Omega))^d \cap V,$$

with the first (Dirichlet) B.C. proper for an eigenfunction, and the second (Neumann) B.C. due to (3.6.21). Our goal now is to show that

$$(3.6.23) \qquad \varphi^* \equiv \sum_{j=1}^{\ell_i} \alpha_j \varphi_{ij}^* \equiv 0 \ \text{in} \ (L^2(\Omega))^d,$$

after which we obtain $\alpha_j \equiv 0$, $j = 1, \ldots, \ell_i$, as desired, since the system $\{\varphi_{ij}^*\}_{j=1}^{\ell_i}$ is linearly independent on $(L^2(\Omega))^d$. To obtain (3.6.23), we proceed as follows. Using the two B.C. for φ^* in (3.6.22), we can extend φ^* by zero locally across all of Γ onto a set ω exterior to Ω, see Fig. 3.5.1 in Section 3.5, to obtain a solution $\tilde{\varphi}^*$ in $(H^1(\tilde{\Omega}))^d$, $\tilde{\Omega} = \Omega \cup \omega$ of the equation in (3.6.22), where $\tilde{\varphi}^* \equiv \varphi^*$ in Ω, and $\tilde{\varphi}^* \equiv 0$ in ω. Thus, by (3.6.22), the extension $\tilde{\varphi}^*$ satisfies equivalently the problem

$$(3.6.24a) \qquad \begin{cases} P[-\nu_0 \Delta \tilde{\varphi}^* + (\tilde{y}_e \cdot \nabla)\tilde{\varphi}^* + (\tilde{\varphi}^* \cdot \nabla)^*\tilde{y}_e)] = \bar{\lambda}_i \tilde{\varphi}^* \ \text{in} \ \tilde{\Omega} = \Omega \cup \omega; \\ \nabla \cdot \tilde{\varphi}^* \equiv 0 \ \text{in} \ \tilde{\Omega}; \ \tilde{\varphi}^* \equiv 0 \ \text{in} \ \partial\tilde{\Omega}, \end{cases}$$
$$(3.6.24b)$$

where $\tilde{y}_e$ is a smooth extension in $\tilde{H} = H_{\tilde{\Omega}}$ of y_e onto ω, where $\tilde{\varphi}^* \equiv 0$ on ω, and $(f \cdot \nabla)^* \tilde{y}_e$ is a d-vector with $\sum_{j=1}^d (D_i \tilde{y}_{ej}) f_j$ as its ith component. Thus, if we eliminate the Leray projector from (3.6.24a), we find that for some $H^1(\tilde{\Omega}))^d$-function $\tilde{p}$, we have:

$$(3.6.25) \qquad -\nu_0 \Delta \tilde{\varphi}^* + (\tilde{y}_e \cdot \nabla)\tilde{\varphi}^* + (\tilde{\varphi}^* \cdot \nabla)^* \tilde{y}_e = \bar{\lambda}_i \tilde{\varphi}^* + \nabla \tilde{p} \ \text{in} \ \tilde{\Omega}.$$

Since $\tilde{\varphi}^* \equiv 0$ in ω, we find ourselves in the same situation as in the proof of [**B-T.1**, Lemma 3.7]. To give one more step, (3.6.25) yields $\nabla \tilde{p} \equiv 0$ in ω, and we may then take $\tilde{p} \equiv 0$ in ω. Hence, we obtain the soughtafter homogeneous four B.C. (see Fig. 3.5.1 in Section 3.5)

$$(3.6.26) \qquad \tilde{p}|_{\partial\tilde{\Omega}} = \partial_\nu \tilde{p}|_{\partial\tilde{\Omega}} = \tilde{\varphi}^*|_{\partial\tilde{\Omega}} = \partial_\nu \tilde{\varphi}^*|_{\partial\tilde{\Omega}} = 0.$$

Then, the proof of [**B-T.1**, Lemma 3.7] applies on $\tilde{\Omega}$ and yields $\tilde{\varphi}^* \equiv 0$, $\tilde{p}^* \equiv 0$ in $\tilde{\Omega}$, hence $\varphi^* \equiv 0$ in Ω, as desired in (3.6.23). The proof of ((3.6.23), hence of) (3.6.20a) is complete. $\square$

PROOF OF (3.6.20B). The proof is the same except for the following changes. Eqn. (3.6.21) has to be interpreted in $(L^2(\Gamma_1))^d$ now, so in (3.6.22) $\partial_\nu \varphi^*|_\Gamma = 0$ is now replaced by $\partial_\nu \varphi^*|_{\Gamma_1} = 0$. Then, φ^* satisfies both Dirichlet and Neumann homogeneous B.C. on Γ_1. Then the set ω is now constructed only across Γ_1, as in Fig. 3.5.2 in Section 3.5. Then (3.6.25) continues to hold true on the new $\tilde{\Omega} = \Omega \cup \omega$ (as in Fig. 3.5.2), with $\varphi^* \equiv 0$ in ω. Then [**B-T.1**, Lemma 3.7] applies and (3.6.20b) is likewise proved.

Step 2. Once the key claim (3.6.20) is established, we see by returning to (3.6.10) that assumption (3.6.16) holds true, at least with (a conservative) $K = N$. In fact, returning to (3.6.10), we see by virtue of property (3.6.20) already proved that: for each $i = 1, \ldots, M$, we can achieve rank $W_i = \ell_i$, for infinitely many choices of the ℓ_i-vectors $[w_{i1}, w_{i2}, \ldots, w_{i\ell_i}]$, where we are thus taking $k = \ell_i$ in (3.6.10). Then, it suffices to take the following $\ell_1 + \ell_2 + \cdots + \ell_M = N$-vectors: $w_1, \ldots, w_N$, defined by

(3.6.27)
$$\underbrace{w_1 = w_{11}, w_2 = w_{12}, \ldots, w_{\ell_1} = w_{1\ell_1}}_{\ell_1}; \underbrace{w_{\ell_1+1} = w_{21}, \ldots, w_{\ell_1+\ell_2} = w_{2\ell_2}}_{\ell_2}, \ldots,$$
$$\underbrace{w_{\ell_1+\cdots+\ell_{M-1}+1} = w_{M1}, \ldots, w_{\ell_1+\cdots+\ell_M} = w_{M,\ell_M}}_{\ell_M},$$

to satisfy all conditions rank $W_i = \ell_i$, $i = 1, \ldots, M$, simultaneously, as then in each corresponding matrix W_i, we can find an $\ell_i \times \ell_i$ minor with vectors $[w_{i1}, \ldots, w_{i\ell_i}]$ with nonzero determinant. Thus, we have proved the statement of Lemma 3.6.2 with $K = N$. □

Finally, to complete the proof of Lemma 3.6.1, we need to show that we can enforce the pointwise c.c. $w_i \cdot \nu = 0$. In fact, in view of definition (3.6.10), the boundary vectors $\{w_i\}_{i=1}^N$ that satisfy hypothesis (3.6.16): rank $W_i = \ell_i$, may be selected to lie in the finite-dimensional subspace $\mathcal{F}$ of $(H^{\frac{1}{2}}(\Gamma))^d$, defined in (3.6.9). Since $\varphi_{ij}^*|_\Gamma \equiv 0$ and $\nabla \cdot \varphi_{ij}^* \equiv 0$ in Ω, then Lemma 3.3.1 applies to φ_{ij}^* and yields that each $\partial_\nu \varphi_{ij}^*$ is orthogonal to the normal ν at each point of Γ. Then, so is w_i, i.e., $w_i \cdot \nu \equiv 0$ on Γ, $i = 1, \ldots, N$, as w_i is a linear combination of the $\partial_\nu \varphi_{ij}^*$. Because of (3.6.20b), the vectors w_i can be chosen to be supported on an arbitrarily small portion Γ_1 of the boundary Γ. □

REMARK 3.6.3. It is possible in many cases to give a far better $K << N$. For instance, if all traces $\{\partial_\nu \varphi_{ij}^*, i = 1, \ldots, M, j = 1, \ldots, \ell_i\}$ happen to be linearly independent on $(L^2(\Gamma))^d$, then we can satisfy assumption (3.6.16) with $K = 1$. Hence, Lemma 3.6.1 and 3.6.2 would hold true with $K = 1$ in this case. Furthermore, if all unstable eigenvalues λ_i of $\mathcal{A}$ happen to have algebraic multiplicity equal to one (simple eigenvalues), then calling $f = \dim \mathcal{F} = \dim[\text{span}\{\partial_\nu \varphi_i^*\}_{i=1}^N]$, assumption (3.6.16) can be satisfied with $K = N - f + 1$, which is then the value of K for Lemmas 3.6.1 and 3.6.2. A more insightful analysis is given elsewhere, for lack of space.

REMARK 3.6.4. Here we pursue the approach in the proof of Lemmas 3.6.1 and 3.6.2 in full generality; that is, in the absence of the FDSA = (3.6.2). Thus, in this Remark, $\{\varphi_{ij}\}_{i=1}^M, {}_{j=1}^{\ell_i}$ are the *generalized* eigenfunctions corresponding to the unstable eigenvalue λ_i of $\mathcal{A}$, with *algebraic* multiplicity ℓ_i. Then, following the treatment in [**B-T.1**], we obtain now that the $\ell_i \times k$ matrices W_i in (3.6.10) and Λ in (3.6.15) have to be replaced with the matrices $\mathcal{W}_i$ and $\tilde{\Lambda}$, respectively,

$$\text{(3.6.28)} \quad \mathcal{W}_i = \begin{bmatrix} (w_1, \partial_\nu(P_{N,i}^*\chi_{i1})), & \ldots, & (w_k, \partial_\nu(P_{N,i}^*\chi_{i1})) \\ (w_1, \partial_\nu(P_{N,i}^*\chi_{i2})), & \ldots, & (w_k, \partial_\nu(P_{N,i}^*\chi_{i2})) \\ \vdots & & \vdots \\ (w_1, \partial_\nu(P_{N,i}^*\chi_{i\ell_i})), & \ldots, & (w_k, \partial_\nu(P_{N,i}^*\chi_{i\ell_i})) \end{bmatrix},$$

$$\tilde{\Lambda} = \left\| \begin{matrix} J_1 & & & 0 \\ & J_2 & & \\ & & \ddots & \\ 0 & & & J_M \end{matrix} \right\|$$

where we have dropped the subscript Γ from the inner product, and where J_i is the Jordan block of dimension ℓ_i corresponding to the distinct unstable eigenvalue λ_i of $\mathcal{A}$, i.e., of $\mathcal{A}_N^u$. Moreover, $\chi_{ir} = \sum_{k=1}^{\ell_i} b_{kr}^{(i)} \varphi_{ik}$, $i = 1, \ldots, M$, for a suitable $\ell_i \times \ell_i$ nonsingular matrix $\mathcal{B}^i \equiv \{b_{kr}^{(i)}\}_{k,r=1}^{\ell_i}$. Accordingly, the counterpart of the key goal in (3.6.20) is now: for each $i = 1, \ldots, M$, show that we have

(3.6.29)   the system $\{\partial_\nu(P_{N,i}^*\chi_{ij})\}_{j=1}^{\ell_1}$ is linearly independent on $(L^2(\Gamma))^d$.

However, since $\{P_{N,i}^*\varphi_{ij}\}_{j=1}^{\ell_i} = M^{(i)}\{\varphi_{ik}^*\}_{k=1}^{\ell_i}$, for $\ell_i \times \ell_i$ nonsingular matrix $M^{(i)}$, proving (3.6.29) is equivalent to proving (3.6.20), this time, however, with $\{\varphi_{ij}^*\}_{j=1}^{\ell_i}$ *generalized* eigenfunctions of $\mathcal{A}^*$ corresponding to the unstable eigenvalue $\bar{\lambda}_i$, with ℓ_i being the *algebraic* multiplicity. This more general version of property (3.6.20) appears not to hold true, qualitatively since a generalized eigenfunction which is not an eigenfunction does not have enough homogeneous B.C. [The two B.C. in (3.6.22) still hold true in the corresponding argument, but now $\mathcal{A}^*\varphi^* = \bar{\lambda}_i\varphi^*$ is replaced by $(\bar{\lambda}_i - \mathcal{A}^*)^{\ell_i}\varphi^* = 0$, a higher-order equation]. Accordingly, it appears that the FDSA cannot be omitted from Lemmas 3.6.1 and 3.6.2.

3.7. Theorem 2.2, case $d = 3$ under the FDSA: An open-loop boundary controller g satisfying the FCC (3.1.22)–(3.1.24) for the linearized system (3.1.4): $g \in L^2(0, \infty; (L^2(\Gamma_1))^d)$, $g \cdot \nu = 0$ on Σ, $g \equiv 0$ on $\Gamma \setminus \Gamma_1$; $y^0 \in (H^{\frac{1}{2}+\epsilon}(\Omega))^d \cap H \Rightarrow y \in L^2(0, \infty; (H^{\frac{3}{2}+2\epsilon}(\Omega))^d \cap H)$

The present section completes the analysis initiated in Section 3.6 under the FDSA. As anticipated in the Orientation of both Section 3.5 and Section 3.6, the next result provides a more desirable class of boundary controllers capable to satisfy the FCC (3.1.22)–(3.1.24) of the linearized system (3.1.4), at the price, however, of requiring the FDSA. Proposition 3.1.3 rules out the possibility of satisfying the FCC (3.1.22)–(3.1.24) of the linearized system (3.1.4), within the class of finite-dimensional controls, for $d = 3$. Accordingly, for $d = 3$, we seek an *infinite-dimensional* (but with a finite-dimensional flavor), open loop boundary controller g, to obtain that the corresponding solution y of the linearized system (3.1.4) satisfies

(3.1.22)–(3.1.24); to wit

(3.7.1) $g \in L^2\left(0, \infty; (L^2(\Gamma))^d\right)$; $g \cdot \nu \equiv 0$ on Γ; $y_0 \in (H^{\frac{1}{2}+\epsilon}(\Omega))^d \cap H$

$$\Rightarrow y \in L^2\left(L^2(0, \infty; (H^{\frac{3}{2}+\epsilon}(\Omega))^d \cap H\right) \text{ continuously.}$$

More precisely, the open-loop controller g here proposed will be a modification of the closed-loop (feedback) controller v_N in Lemma 3.6.1, see (3.6.12):
(3.7.2)
$$v_N(t) = \sum_{i=1}^K (e^{\bar{A}^u t} P_N z_0, p_i)_H w_i \in C^n(0, \infty; \mathcal{F}); \quad n = 1, 2, \ldots, \quad P_N z_0 = z_N(0),$$

so far considered for $d = 2, 3$, yielding feedback stabilization of the finite-dimensional z_N-problem (3.4.9) [as well as stabilization of the full linearized system (3.1.4), however, only in the $(H^{\frac{3}{2}-2\epsilon}(\Omega))^d$-topology, see Proposition B.2.1 to follow in Appendix B.2, which is sufficient also in the non-linear case (1.1) for $d = 2$, but not for $d = 3$]. We first prove Theorem 2.2 in the complexified case.

PROPOSITION 3.7.1. *Assume the FDSA = (3.6.2). We consider the complexified z-problem (3.4.1) and its projected versions: the z_N-problem in (3.4.9) and the ζ_N-problem in (3.4.10). We assume the rank condition (3.6.16): rank $W_i = \ell_i$, $i = 1, \ldots, M$, where W_i is defined in the present diagonalizable (semi-simple) case by (3.6.10) for suitable boundary vectors $w_1, \ldots, w_K \in \mathcal{F} = \text{span}\{\partial_\nu \varphi^*_{ij}\}_{i=1}^{M}{}_{j=1}^{\ell_i}$, $K = N$ (conservatively). We can require that $w_i \cdot \nu \equiv 0$ on Γ, and that the w_i's be supported on any small portion Γ_1 of the boundary Γ, of finite measure (as in Lemma 3.6.1), the latter property because of (3.6.20b).*

Let $p_1, \ldots, p_K$ be the vectors on Z_N^u provided by Lemma 3.6.2 in correspondence of the arbitrarily preassigned number γ_1, so that

(3.7.3) $\qquad |z_N(t)| = \left|e^{\bar{A}^u t} z_N(0)\right| \leq C_{\gamma_1} e^{-\gamma_1 t} |z_N(0)|, \ t \geq 0, \ z_N(0) = P_N z_0,$

in the notation of (3.6.11), (3.6.18). Introduce the open-loop boundary control

(3.7.4) $\quad g(t; z_0) = g(t) = \sum_{i=1}^K (e^{\bar{A}^u t} P_N z_0, p_i)_H w_i + \mathfrak{g}_1(t)\mathfrak{g}_2(x)$

$\qquad \in H^n(0, \infty; (L^2(\Gamma))^d), \ g \cdot \nu \equiv 0 \text{ on } \Sigma, \ n = 1, 2, \ldots,$

where the first closed-loop $\sum_{i=1}^K$-term is precisely provided by Lemma 3.6.2, see the RHS of (3.6.12) recalled in (3.7.2); while $\mathfrak{g}_1(t)$ and $\mathfrak{g}_2(x)$ in the second open-loop term, are a time-dependent and a boundary space-dependent function, respectively, defined by

(3.7.5) $\qquad \mathfrak{g}_1(t) = e^{-2\gamma_1 t}; \quad \mathfrak{g}_2 = z_0|_\Gamma - \sum_{i=1}^K (P_N z_0, p_i)_H w_i \in (L^2(\Gamma))^d$

(thus $\mathfrak{g}_2$ depends on the I.C. z_0). Moreover, since $w_i \cdot \nu = 0$ on Γ by virtue of Lemma 3.3.1 (see statement of Lemma 3.6.1 above (3.6.11)) and since $z_0|_\Gamma \cdot \nu = 0$ on Γ (as $z_0 \in H$), we likewise have: $\mathfrak{g}_2 \cdot \nu = 0$ on Γ.

Then: the boundary control g in (3.7.4) satisfies the following properties:

3.7. THEOREM 2.2, CASE $d = 3$ UNDER THE FDSA

(i) g satisfies the boundary compatibility condition; see (3.1.26):

$$(3.7.6) \quad (z|_\Gamma)_{t=0} = z_0|_\Gamma = g|_{t=0} = g(0) = \sum_{i=1}^{K}(P_N z_0, p_i)w_i + \mathfrak{g}_2 \in (L^2(\Gamma))^d;$$

(ii) such g given by (3.7.4), once inserted in model (3.4.1) with $v = g$, yields

$$(3.7.7) \quad \frac{dz}{dt} - \mathcal{A}z = -\mathcal{A}D\left[\sum_{i=1}^{K}(e^{\bar{A}_u t}P_N z_0, p_i)w_i + \mathfrak{g}_1(t)\mathfrak{g}_2(x)\right], \quad z(0) = z_0$$

and produces a solution $z(t)$ satisfying the following two properties
(ii_1) exponential decay in H:

$$(3.7.8) \quad |A^\theta z(t)| \leq C_{\gamma_0,\delta,\theta} e^{-\gamma_0 t}|z_0|, \quad t \geq \delta > 0, \ 0 \leq \theta < \frac{1}{4},$$

for any constant γ_0 satisfying $0 < \gamma_0 < |\mathrm{Re}\ \lambda_{N+1}|$, where we can take $\delta = 0$ for $\theta = 0$;
(ii_2) if $z_0 \in (H^{\frac{1}{2}+2\epsilon}(\Omega))^d \cap H$, then $[z_0 - Dg(0)] \in \mathcal{D}((-\mathcal{A})^{\frac{1}{4}+\epsilon})$

$$(3.7.9a) \quad \begin{cases} z \in C\left([0,T]; (H^{\frac{1}{2}+2\epsilon}(\Omega))^d \cap H\right) \cap L^2\left(0,\infty; (H^{\frac{3}{2}+2\epsilon}(\Omega))^d \cap H\right); \\ z_t \in L^2\left(0,\infty; (H^{-\frac{1}{2}+2\epsilon}(\Omega))^d \cap H\right); \\ z|_\Gamma = g \in C([0,T]; (H^{2\epsilon}(\Gamma))^d \cap H) \cap L^2(0,\infty; (L^2(\Gamma))^d). \end{cases}$$
(3.7.9b)
(3.7.9c)

PROOF. (i) Verification of (3.7.6) is immediate from the definition of $g(t)$ in (3.7.4), with $z_N(0) = P_N z_0$, $\mathfrak{g}_1(0) = 1$.

(ii_1) Eqn. (3.7.8) is the same as [Proposition B.2.1, Appendix B.2, Eqn. (B.2.5)] or [**B-T.1**].

(ii_2) Let $z_0 \in (H^{\frac{1}{2}+2\epsilon}(\Omega))^d \cap H$. Then, via (3.7.6) just verified, we have $z_0|_\Gamma = g(0) \in (H^{2\epsilon}(\Omega))^d$ by trace theory; hence $Dg(0) \in (H^{\frac{1}{2}+2\epsilon}(\Omega))^d \cap H$ by (3.1.3d). Thus, $[z_0 - Dg(0)] \in (H_0^{\frac{1}{2}+2\epsilon}(\Omega))^d \cap H = \mathcal{D}((-\mathcal{A})^{\frac{1}{4}+\epsilon})$ by (1.15). Moreover, $g_t \in L^2(0,\infty; H)$, see (3.7.4). Then, Lemma 3.1.2(ii) applies, and (3.7.9) then is (the complex version of) (3.1.18a-c). $\square$

REMARK 3.7.1. The boundary control $g(t; z_0)$ in (3.7.4) has a few desirable features:
(a)

$$(3.7.10) \quad g(t; z_0) - e^{-2\gamma_1 t}(z_0|_\Gamma)$$

$$= \sum_{i=1}^{K}\left(e^{\bar{A}_u t}P_N z_0, p_i\right)_H w_i - e^{-2\gamma_1 t}\sum_{i=1}^{K}(P_N z_0, p_i)_H w_i \in (H^{\frac{1}{2}}(\Gamma))^d$$

$$= \text{finite-dimensional}$$

(b) The boundary vectors w_i may be required to have support only on an arbitrarily small part Γ_1 of Γ (of positive surface measure), by virtue of property (3.6.20b).

A remark such as Remark 5.1 applies now.

Completion of the proof of Theorem 2.2. Apply Proposition 3.7.1 and take the real part of the complexified system (3.4.1), with control $v = g$ given by

(3.7.4). We get, recalling $z_N(t) = e^{\bar{A}^u t} P_N z_0$, see below (3.6.12):

(3.7.11) $$y_t - \mathcal{A}y = -\mathcal{A}D(\text{Re } g),$$

(3.7.12) $$\text{Re } g = \sum_{i=1}^{K} \text{Re}[(z_N(t), p_i)_H w_i] + e^{-2\gamma_1 t} y_0|_\Gamma$$
$$- e^{-2\gamma_1 t} \sum_{i=1}^{K} \text{Re}[(P_N z_0, p_i)_H w_i].$$

Setting for $i = 1, \ldots, K$,

(3.7.13) $$u_N^i(t) = \text{Re}(z_N(t), p_i)_H; \qquad u_N^{i+K}(t) = -\text{Im}(z_N(t), p_i)_H;$$

(3.7.14) $$\tilde{p}_i = \text{Re } p_i - \text{Im } p_i; \qquad \tilde{w}_i = \text{Re } w_i, \ \tilde{w}_{i+K} = \text{Im } w_i,$$

and selecting $P_N z_0 = P_N y^0 + i P_N y^0$, we obtain from (3.7.11)–(3.7.14):

(3.7.15) $$\text{Re } g = \sum_{i=1}^{2K} u_N^i(t) \tilde{w}_i + e^{-2\gamma_1 t} y^0|_\Gamma$$
$$- e^{-2\gamma_1 t} \sum_{i=1}^{K} (P_N y^0, \tilde{p}_i)_H (\tilde{w}_i - \tilde{w}_{i+K}),$$

which is (2.5). □

CHAPTER 4

Boundary feedback uniform stabilization of the linearized system (3.1.4) via an optimal control problem and corresponding Riccati theory. Case $d = 3$

4.0. Orientation

The goal of the present section is to introduce and develop at a somewhat abstract level an optimal control problem over an infinite time-horizon with a quadratic cost involving a 'highly unbounded' observation operator—essentially by a critical "2ϵ" (see below) beyond the abstract theory available in the literature [**L-T.1**], [**B-T.1**]—such as it occurs in the case of the linearized Navier-Stokes system (3.1.4) for $d = 3$. This is the result of the analysis given in Section 3.5. As usual, a key benefit of an optimal control theory and corresponding feedback Riccati theory when it works—and here the challenge is to make it work with the high degree of unboundedness of the observation operator—is a *uniform stabilization* result of the feedback semigroup, as an immediate corollary (via Datko's result [**P.1**, p. 116]) of the existence of an optimal solution. Thus, when tuned to the case of the linearized Navier-Stokes system (3.1.4), this optimal control theory over an infinite time-horizon provides—as a by-pass product—the *uniform stabilization of the linearized problem* (3.1.4), $d = 3$ (*a-fortiori*, $d = 2$) by virtue of a boundary feedback controller constructed by means of a suitable Riccati operator. With reference to the linearized Navier-Stokes problem (3.1.4), in concrete terms, the aforementioned statement "by a critical 2ϵ beyond the abstract theory available in the literature" means the passage from the $H^{\frac{3}{2}-\epsilon}(\Omega)$-topology [**L-T.1**], [**L-T.2**] (or the $\mathcal{D}(A^{\frac{3}{4}})$-topology [**B-T.1**]) to the $H^{\frac{3}{2}+\epsilon}(\Omega)$-topology for the observation space. More precisely: the case of $H^{\frac{3}{2}-\epsilon}(\Omega)$ as an observation space for the linearized system (3.1.4) occurs for $d = 2$, which is studied in full generality in Appendix A, whose Riccati theory is covered by available literature [**L-T.1**], [**L-T.2**], [**B-T.1**]. Instead, the case of $H^{\frac{3}{2}+\epsilon}(\Omega)$ as an observation space for the linearized system (3.1.4) occurs for $d = 3$, and was studied in Section 3.5 in full generality and Section 3.7 under the FDSA. The key result of these two sections—upon which the present Section 4 rests critically—is the statement that the optimal control problem is well-posed with $(H^{\frac{3}{2}+\epsilon}(\Omega))^d \cap H$, $d = 3$, as observation space; more precisely, *that the Finite Cost Condition (3.1.22)–(3.1.25) holds true for the linearized system (3.1.4)*: see Theorem 3.5.1 in full generality; and Theorem 3.7.1 under the FDSA. Under this preliminary critical property, the study of the optimal control problem may then begin. In the present case ($d = 3$, $(H^{\frac{3}{2}+2\epsilon}(\Omega))^3 \cap H$ as an observation space), we need to develop the present new treatment of the optimal control problem and related Riccati theory, a technical and conceptual extension of [**F-L-T.1**], [**L-T.1**], [**L-T.2**], [**B-T.1**]. We close this Orientation by pointing out that, due to the 'high degree of unboundedness' of the observation operator, the Algebraic Riccati Equation for the feedback solution of the optimal control problem is asserted to hold true only on the domain of the feedback generator—not of the original free dynamics generator. This more limited conclusion is sufficient, however, to achieve the sought-after uniform stabilization of the feedback semigroup, corresponding to the optimal dynamics.

REMARK 4.0.1. Finally, the specific degree of unboundedness here taken for the observation operator—that is the particular high-level topology taken for the observation space—is dictated by the non-linear term $(y \cdot \nabla)y$ of the Navier-Stokes equation, via Sobolev's embedding in the case $d = 3$: see Eqn. (5.18b) below. Concretely, this leads in this case with $d = 3$ to the observation space $(H^{\frac{3}{2}+\epsilon}(\Omega))^3 \cap H$, in order for the observation operator to be sufficiently unbounded to control the non-linear term.

If, however, one is interested only in the observation space $(H^{\frac{3}{2}-\epsilon}(\Omega))^d \cap H$, $d = 2, 3$, for the linearized problem (3.1.4)—which only in the case $d = 2$ is sufficient for the stabilization of the full Navier-Stokes equation (1.1)—then the validity of the corresponding FCC is asserted as follows. For $d = 2$, in the general case, in Appendix B.1 with an *infinite-dimensional* boundary controller acting on an *arbitrarily small portion* of the boundary.

For $d = 2, 3$, under the FDSA = (3.6.2): in Appendix B.2 with a *finite-dimensional* boundary controller acting on an *arbitrarily small portion* of the boundary.

4.1. The optimal control problem (Case $d = 3$)

Spaces and operators. Throughout the present Section 4, we shall use the following Hilbert spaces: U (control space); W (state space of the *optimal feedback dynamics*), H (state space of the *free dynamics*); Z (observation space), where:

(4.1.1a) $\qquad U \equiv \{u \in (L^2(\Gamma))^d, u \cdot \nu = 0 \text{ on } \Gamma\};$

$\qquad H \equiv$ space defined in (1.5);

(4.1.1b) $\qquad W \equiv \left(H^{\frac{1}{2}+\epsilon}(\Omega)\right)^d \cap H;$

$\qquad Z \equiv \left(H^{\frac{3}{2}+\epsilon}(\Omega)\right)^d \cap H,$

with $d = 3$ (demanding case), where $\epsilon > 0$ arbitrary but fixed, once and for all. Moreover, $\mathcal{A}$, D are the operators in (1.11) and (3.1.3) respectively, while

(4.1.1c) $\qquad Q$ (observation operator): canonical isomorphism Z onto H.

Dynamics. We return to the abstract model (3.1.4a) of the linearized problem (3.1.2)

(4.1.2) $\qquad y' - \mathcal{A}y = -\mathcal{A}Du \in [\mathcal{D}(\mathcal{A}^*)]', \ t > 0;$

$\qquad y(0) = y_0 \in W \equiv \left(H^{\frac{1}{2}+\epsilon}(\Omega)\right)^3 \cap H,$

where $Du \in H$, since $u \cdot \nu \equiv 0$, see U in (4.1.1a). Thus, $\mathcal{B} = -\mathcal{A}D$ in the usual notation.

Optimal control problem on $[0, \infty]$. With the dynamics (4.1.2), we associate the quadratic cost functional

$$(4.1.3) \qquad J(u,y) \equiv \int_0^\infty [|Qy(t)|_H^2 + |u(t)|_U^2]dt, \quad \text{where } |Qy|_H = |y|_Z,$$

where $y(t) = y(t; y_0; u)$ is the solution of Eqn. (4.1.2) due to $u \in L^2(0, \infty; U)$. In this section we use y_0 instead of y^0. The optimal control problem is

$$(4.1.4) \qquad \text{Minimize } J(u,y) \text{ over all } u \in L^2(0, \infty; U).$$

We refer to Remark 4.1, and Remark 5.1 below, for a justification of the above setting, for $d = 3$, with cost penalization of y at the $(H^{\frac{3}{2}+\epsilon}(\Omega))^d$-level. The treatment below conforms to the usual formula of the literature [**F-L-T.1**], [**L-T.1**], [**L-T.2**], with, formally, $\mathcal{B} = -\mathcal{A}D$, $\mathcal{B}^* = -D^*\mathcal{A}^*$.

As mentioned in the Orientation, our first step consists in establishing that this optimal control problem has a unique solution. To this end, it suffices to assert that the Finite Cost Condition (FCC) holds true. This is a challenging issue for $d = 3$, which has been resolved in Section 3.5.

PROPOSITION 4.1.1. *With reference to (4.1.1)–(4.1.4), for each given $y_0 \in W$, there exists some open-loop boundary control $u \in L^2(0, \infty; U)$ such that, along with the corresponding solution y of (4.1.2), it satisfies $J(u, y) < \infty$.*

PROOF. This is asserted for $d = 3$ in full generality in Theorem 3.5.1 of Section 3.5 and under the FDSA = (3.6.2) in Proposition 3.7.1 of Section 3.7. □

The following consequence of Proposition 4.1.1 follows in a standard way.

COROLLARY 4.1.2. *(i) For each $y_0 \in W$, there exists a unique optimal solution pair $\{u^*(\,\cdot\,; y_0), y^*(\,\cdot\,; y_0)\}$, $u^* \cdot \nu = 0$ on Γ, of the optimal control problem (4.1.3), (4.1.4):*

$$(4.1.5a) \qquad u^*(\,\cdot\,; y_0) \in L^2(0, \infty; U); \quad y^*(\,\cdot\,; y_0) \in L^2(0, \infty; Z).$$

$$(4.1.5b) \qquad Qy^*(\,\cdot\,; y_0) \in L^2(0, \infty; H).$$

(ii) There exists a non-negative, self-adjoint bounded operator $\hat{R}$ on W, $\hat{R} \in \mathcal{L}(W)$, such that, for all $y_0 \in W$:

$$(4.1.6) \qquad (\hat{R}y_0, y_0)_W = \min_{u \in L^2(0,\infty;U)} J(u, y) \equiv J(u^*(\,\cdot\,; y_0), y^*(\,\cdot\,; y_0)) \equiv J^*(y_0).$$

(iii) In view of the duality pairing,

$$(4.1.7) \qquad W \subset H = H' \subset W',$$

and rescaling, there exists a non-negative, self-adjoint operator R on H, such that

$$(4.1.8a) \qquad R \in \mathcal{L}(W; W'); \quad J^*(x) = (Rx, x)_H = (\hat{R}x, x)_W \leq C|x|_W^2, \quad x \in W;$$

$$(4.1.8b) \qquad \int_0^\infty (Qy^*(t, 0; x), Qy^*(t, 0; y))_H + (u^*(t, 0; x), u^*(t, 0; y))_U \, dt$$
$$= (Rx, y)_H, \quad x, y \in W.$$

We next complement (4.1.5) with optimal regularity properties.

PROPOSITION 4.1.3. *The optimal pair in (4.1.5) for $d = 3$ satisfies the following regularity properties for $y_0 \in W$, and any $0 < T < \infty$:*

(4.1.9a) $\quad y^*(\,\cdot\,;y_0) \in L^2(0,\infty;(H^{\frac{3}{2}+\epsilon}(\Omega))^d \cap H) \cap C([0,T];(H^{\frac{1}{2}+\epsilon}(\Omega))^d \cap H)$

(4.1.9b) $\quad y_t^*(\,\cdot\,;y_0) \in L^2(0,\infty;(H^{-\frac{1}{2}+\epsilon}(\Omega))^d);$

(4.1.10) $\quad y^*(\,\cdot\,;y_0)|_\Gamma \equiv u^*(\,\cdot\,;y_0) \in L^2(0,\infty;(H^{1+\epsilon}(\Gamma))^d) \cap C([0,T];(H^\epsilon(\Gamma))^d).$

PROOF. Starting with (4.1.5), we proceed as in (3.1.22)–(3.1.25):

(4.1.11) $\quad y^*(\,\cdot\,;y_0) \in L^2\left(0,\infty;(H^{\frac{3}{2}+\epsilon}(\Omega))^d \cap H\right)$

$\quad \Rightarrow y_t^*(\,\cdot\,;y_0) \in L^2\left(0,\infty;(H^{-\frac{1}{2}+\epsilon}(\Omega))^d \cap H\right),$

since $y_t^* = -P\mathbb{A}y^*$ in $\Omega \times (0,\infty)$, where $Py_t^* = y_t^*$ as $y^* \in H$. We then apply to (4.1.11) [**L-M.1**, Theorem 3.1, p. 19] to get

(4.1.12) $\quad y^*(\,\cdot\,;y_0) \in C([0,T];(H^{\frac{1}{2}+\epsilon}(\Omega))^d \cap H),$

since $[(H^{\frac{3}{2}+\epsilon}(\Omega))^d, (H^{-\frac{1}{2}+\epsilon}(\Omega))^d]_{\frac{1}{2}} = (H^{\frac{1}{2}+\epsilon}(\Omega))^d$. Thus (4.1.11) and (4.1.12) yield (4.1.9). Then, trace theory on y^* in (4.1.11) and y^* in (4.1.12) yields (4.1.10) for $y^*|_\Gamma = u^*$. □

Our next result shows that the operator R of Corollary 4.1.2(iii) is coercive.

PROPOSITION 4.1.4. *With reference to the non-negative, self-adjoint operator R in (4.1.8) on H, there exists a constant $c > 0$, such that*

(4.1.13a) $\quad J^*(x) = (Rx,x)_H = |R^{\frac{1}{2}}x|_H^2 \geq c|x|_W^2, \quad \forall\, x \in W,$

so that combining with (4.1.8), we get the equivalence relation

(4.1.13b) $\quad c|x|_W^2 \leq J^*(x) = (Rx,x)_H \leq C|x|_W^2, \quad \forall\, x \in W$

PROOF. The equality on the LHS of (4.1.13a) was already stated in (4.1.8). To show the inequality in (4.1.13a) (coercivity), we consider the optimal control problem (4.1.3), (4.1.4) with an as yet unspecified initial condition y_0, and we assume that there exists a (unique) optimal solution $\{u^*(\,\cdot\,;y_0), y^*(\,\cdot\,;y_0)\}$, so that $J^*(y_0) < \infty$. We then seek to show that, in fact, $y_0 \in W$: that is, $J^*(y_0) < \infty \Rightarrow y_0 \in W$. After that, we then obtain that the map $J^*(y_0) \to y_0 \in W$ is continuous, and hence inequality (4.1.13a) follows. To this end, we invoke (4.1.9) for $y^*(\,\cdot\,;y_0)$, set $t = 0$, and get $y^*(0;y_0) = y_0 \in (H^{\frac{1}{2}+\epsilon}(\Omega))^d \cap H \equiv W$, as desired. □

4.2. Optimal feedback dynamics: the feedback semigroup and its generator on W

Optimal control problem on $[s,\infty]$, $s \geq 0$. We now consider the optimal control problem (4.1.3), (4.1.4) on the time interval $[s,\infty]$, rather than $[0,\infty]$, with initial condition $y(s) = y_s \in W$ at the initial time $t = s$, under the corresponding FCC guaranteed by Proposition 4.1.1. We shall accordingly denote by $\{u^*(t,s;y_s), y^*(t,s;y_s)\}$, $s \leq t \leq \infty$, the corresponding unique optimal pair (Corollary 4.1.2(i)). The regularity result corresponding to Proposition 4.1.3 is then

4.2. OPTIMAL FEEDBACK DYNAMICS

PROPOSITION 4.2.1. *For each $y_s \in W$, the aforementioned unique optimal pair $\{u^*(t, s; y_s), y^*(t, s; y_s)\}$ of the optimal control problem (4.1.3), (4.1.4) except on $[s, \infty]$ rather than on $[0, \infty]$ with initial datum $y_s \in W \equiv (H^{\frac{1}{2}+\epsilon}(\Omega))^d \cap H$, at the initial time $t = s$, satisfies the following regularity properties (recalling (4.1.1b)) for any $0 \leq s < T < \infty$:*

(4.2.1) $\quad y^*(\,\cdot\,, s; y_s) \in C([s, T]; W) \cap H^\theta\left(s, \infty; (H^{\frac{3}{2}+\epsilon-2\theta}(\Omega))^d \cap H\right), \quad 0 \leq \theta \leq 1;$

(where "$\cap H$" is omitted for $\frac{3}{2} + \epsilon - 2\theta < 0$ or $\theta > \frac{3}{4} + \frac{\epsilon}{2}$)

(4.2.2) $\quad u^*(\,\cdot\,, s; y_s) \in C([s, T]; (H^\epsilon(\Gamma))^d) \cap H^\theta\left(s, T; (H^{1+\epsilon-2\theta}(\Gamma))^d\right),$

$$0 \leq \theta < \frac{1}{2} + \frac{\epsilon}{2}.$$

PROOF. The regularity (4.2.1) (right) follows by interpolation between (4.1.9a) and (4.1.9b). Then this result implies (4.2.2) by trace theory with $\frac{3}{2}+\epsilon-2\theta > \frac{1}{2}$. □

The feedback semigroup describing the optimal evolution. Based on the regularity properties of the optimal solution in Proposition 4.2.1, we are in a position to define an operator $S(t) \in \mathcal{L}(W)$ by

(4.2.3) $\quad S(t)x \equiv y^*(t, 0; x) \in C([0, T]; W), \quad x \in W.$

PROPOSITION 4.2.2. *(i) The operator $S(t)$, $0 \leq t < \infty$, in (4.2.3) is a strongly continuous semigroup on W.*

(ii) Moreover, $S(t)$ is (exponentially) uniformly stable on W: there exist constants: $M \geq 1$, $a > 0$, such that

(4.2.4) $\quad |S(t)x|_W \leq M e^{-at} |x|_W, \quad \forall\, x \in W.$

PROOF. (i) Strong continuity and $S(0)x = x$ is plain. We need to establish the semigroup property; that is, that for $0 \leq s \leq t$:

(4.2.5) $\quad y^*(t, 0; y_0) = y^*(t, s; y^*(s, 0; y_0)) = y^*(t - s, 0; y^*(s, 0; y_0));$

after which (4.2.3) gives: $S(t)y_0 = S(t-s)S(s)y_0$, as desired. But the identities in (4.2.5), along with the companion identities

(4.2.6) $\quad u^*(t, 0; y_0) = u^*(t, s; y^*(s, 0; y_0)) = u^*(t - s, 0; y^*(s, 0; y_0));$

follow pairwise from the *optimization* problem, based on the notion that: given the optimal solution pair $\{u^*(t, s; y_s), y^*(t, s; y_s)\}$ over the time interval $[s, \infty]$, originating at the point y_s at the time $t = s$, then the pair

$$\{u^*(t, \tau; y^*(\tau, s; y_s)), y^*(t, \tau; y^*(\tau, s; y_s))\}$$

is a compatible pair (the first term produces the second the equation (4.1.2)) over the time interval $[\tau, \infty]$, $s \leq \tau$, originating at the point $y^*(\tau, s; y_s)$ of the preceding optimal trajectory. Hence this latter pair is a competing pair for the optimization problem over $[\tau, \infty]$, which then readily turns out to be the (unique) optimal pair on $[\tau, \infty]$, with $y^*(\tau, s; y_s)$ as the initial point. This, in fact, yields that for any $0 \leq s < t$, we have

(4.2.7) $\quad \displaystyle\int_s^\infty [|y^*(t, 0; y_0)|_Z^2 + |u^*(t, 0; y_0)|_U^2]\,dt$

$\qquad = \displaystyle\int_s^\infty [|y^*(t, s; y^*(s, 0; y_0))|_Z^2 + u^*(t, s; y^*(s, 0; y_0))|_U^2]\,dt,$

from which we obtain the top identity in (4.2.5) (left) and (4.2.6) (left), first a.e., and then pointwise via (4.2.3). Then, the right identities in (4.2.5), (4.2.6) follow, as the equation is autonomous.

(ii) The optimization problem yields *a-fortiori* for all $y_0 \in W$, recalling (4.1.1b):

$$(4.2.8) \quad \int_0^\infty |S(t)y_0|_W^2 dt \leq \int_0^\infty |S(t)y_0|_Z^2 dt = \int_0^\infty |y^*(t,0;y_0)|_Z^2 dt \leq C|y_0|_W^2,$$

where in the last step we have invoked (4.1.8). Then, a well-known result by Datko [**P.1**, p. 116] applies and yields (4.2.4). □

Henceforth, let $A_R : W \supset \mathcal{D}(A_R) \to W$ denote the (closed, densely defined) generator of the s.c. semigroup $S(t)$: $S(t) = e^{A_R t}$, $t \geq 0$, on W:

$$(4.2.9) \quad \begin{cases} \dfrac{dS(t)x}{dt} = A_R S(t)x = S(t)A_R x, \quad x \in \mathcal{D}(A_R); \\ S(t)x = e^{A_R t}x, \quad x \in W, \text{ so that } A_R^{-1} \in \mathcal{L}(W) \text{ from (4.2.4).} \end{cases}$$

We shall next establish some regularity properties of the domain $\mathcal{D}(A_R)$ of A_R.

PROPOSITION 4.2.3. *With reference to the generator A_R of $S(t)$ on W and the observation operator Q in (4.1.1c), we have:*

(i) $\mathcal{D}(A_R) \subset \mathcal{D}(Q) = (H^{\frac{3}{2}+\epsilon}(\Omega))^d \cap H$; *i.e.,* $x \in \mathcal{D}(A_R) \Rightarrow Qx \in H$ *continuously:*

$$(4.2.10) \quad |Qx|_H \leq C|A_R x|_W, \quad \forall\, x \in \mathcal{D}(A_R) \text{ or } QA_R^{-1} \in \mathcal{L}(H);$$

(ii) generalizing (4.2.10) with $Z = (H^{\frac{3}{2}+\epsilon}(\Omega))^d \cap H$,

$$(4.2.11) \quad |QS(t)x|_H = |S(t)x|_Z \leq Ce^{-at}|A_R x|_W, \ t \geq 0, \quad \forall\, x \in \mathcal{D}(A_R);$$

(iii) by trace theory

$$(4.2.12) \quad |u^*(t,0;x)|_{(H^{1+\epsilon}(\Gamma))^d} \leq Me^{-at}|A_R x|_W, \quad \text{a.e. in } t, \quad \forall\, x \in \mathcal{D}(A_R).$$

PROOF. (i), (ii) We recall the definition (4.1.8) of the self-adjoint operator R on H:

$$(4.2.13) \quad (Rx,x)_H = J^*(x) = \int_0^\infty [|Qy^*(\tau,0;x)|_H^2 + |u^*(\tau,0;x)|_U^2]d\tau$$

$$(\text{by (4.2.3)}) \quad = \int_0^\infty [|QS(\tau)x|_H^2 + |u^*(\tau,0;x)|_U^2]d\tau, \ x \in W.$$

Thus, replacing $x \in W$ with $y^*(t,0;x) = S(t)x \in W$, see (4.2.3), and recalling that $u^*(\tau,0;S(t)x) = u^*(\tau,0;y^*(t,0;x)) = u^*(\tau+t,0;x)$ by (4.2.6) with $s=t$, $t-s=\tau$, we obtain from (4.2.13) and the semigroup property:

$$(4.2.14) \quad (RS(t)x, S(t)x)_H = \int_0^\infty [|QS(\tau+t)x|_H^2 + |u^*(t+\tau;0;x)|_U^2]d\tau$$

$$(t+\tau = s) \quad = \int_t^\infty [|QS(s)x|_H^2 + |u^*(s,0;x)|_U^2]ds, \ x \in W.$$

Next, we specialize (4.2.14) to $x \in \mathcal{D}(A_R)$, differentiate (4.2.14) in t, recall (4.2.9), and obtain since $R: W \to W'$ is self-adjoint on H (Corollary 4.1.2(iii)):

$$|QS(t)x|_H^2 + |u^*(t,0;x)|_U^2$$

(4.2.15)
$$= -(RS(t)A_R x, S(t)x)_H$$
$$\quad -(RS(t)x, S(t)A_R x)_H, \quad x \in \mathcal{D}(A_R)$$

(4.2.16) $\quad \leq \quad 2|RS(t)A_R x|_{W'}|S(t)x|_W \leq c|S(t)A_R x|_W |S(t)x|_W$

(4.2.17) (by (4.2.4)) $\quad \leq \quad Ce^{-2at}|A_R x|_W |x|_W \leq Ce^{-2at}|A_R x|_W^2, \quad x \in \mathcal{D}(A_R).$

In going from (4.2.16) to (4.2.17) we have invoked (4.2.4) for $S(t)$, hence $A_R^{-1} \in \mathcal{L}(W)$, so that $|x|_W \leq c|A_R x|_W$, a property used in the last step of (4.2.17). Thus, (4.2.17) proves (4.2.11). Setting $t = 0$ in (4.2.11) or (4.2.17) proves (4.2.10).

(iii) We use (4.1.10), trace theory, and (4.2.11), for $x \in \mathcal{D}(A_R)$:

(4.2.18) $\quad |u^*(t,0;x)|_{(H^{1+\epsilon}(\Gamma))^d}$
$$= |y^*(t,0;x)|_\Gamma|_{(H^{1+\epsilon}(\Gamma))^d} \leq c|y^*(t,0;x)|_{(H^{\frac{3}{2}+\epsilon}(\Omega))^d \cap H = Z}$$

(4.2.19) $\quad = c|QS(t)x|_H \leq Ce^{-at}|A_R x|, \quad x \in \mathcal{D}(A_R), \quad \text{a.e. in } t,$

and (4.2.12) is established. In going from (4.2.18) to (4.2.19), we have invoked (4.1.1c) for Q and (4.2.3), and in (4.2.19) we have used (4.2.11). $\square$

4.3. Feedback synthesis via the Riccati operator

At this point, we make reference to Appendix C; in particular, to Theorem C.1, giving a desirable *equivalence* between the present OCP (4.1.3), (4.1.4)—where the free dynamics semigroup $e^{\mathcal{A}t}$ is unstable [OCP#1, (C.1), (C.2), in Appendix C]—and its counterpart where the generator $\mathcal{A}$ is suitably translated into $(\mathcal{A} - \lambda)$, $\lambda > 0$, so that the corresponding semigroup $e^{(\mathcal{A}-\lambda)t}$ is exponentially stable [OCP#2, (C.5), (C.6) in Appendix C]. Thus, henceforth, in light of Theorem C.1, we shall assume *without loss of generality* for our study of the OCP (4.1.2), (4.1.3), that: the original semigroup $e^{\mathcal{A}t}$ is exponentially stable on H: there exist constants $C \geq 1$, $\delta > 0$, such that

(4.3.0) $\qquad |e^{\mathcal{A}t}|_{\mathcal{L}(H)} \leq Ce^{-\delta t}, \qquad t \geq 0.$

Taking advantage of (4.3.0) and motivated by available theory [**F-L-T.1**], [**L-T.1**], we can introduce the following operator R_1:

(4.3.1) $\qquad R_1 x \equiv \int_0^\infty e^{\mathcal{A}^* t} Q^* Q S(t) x \, dt = \int_t^\infty e^{\mathcal{A}^*(\tau - t)} Q^* Q S(\tau - t) x \, d\tau,$

initially for, say, $x \in \mathcal{D}(A_R)$; eventually for

$$x \in W \equiv \left(H^{\frac{1}{2}+\epsilon}(\Omega) \right)^d \cap H,$$

whose regularity—a key issue of the present analysis—will be discussed below. In preparation for this, we recall a few facts: (i) First that $Qy^*(t,0;x) \equiv QS(t)x \in L^2(0,\infty; H)$ for $x \in W$, by well-posedness of the OCP in (4.1.5); (ii) next, that

$\mathcal{D}((-\mathcal{A})^{\frac{3}{4}+\frac{\epsilon}{2}}) \subset (H^{\frac{3}{2}+\epsilon}(\Omega))^d \cap H$, see (1.17); (iii) finally, that Q: continuous $(H^{\frac{3}{2}+\epsilon}(\Omega))^d \cap H \to H$, as in (4.1.1c). Then, (ii) and (iii) combined imply that:

(4.3.2a) $\quad \begin{cases} Q^* : \text{ continuous } H \to \left[(H^{\frac{3}{2}+\epsilon}(\Omega))^d \cap H\right]' \subset \left[\mathcal{D}((-\mathcal{A})^{\frac{3}{4}+\frac{\epsilon}{2}})\right]'; \\ \text{equivalently, } (-\mathcal{A}^*)^{-(\frac{3}{4}+\frac{\epsilon}{2})}Q^* \in \mathcal{L}(H), \text{ or } Q(-\mathcal{A})^{-(\frac{3}{4}+\frac{\epsilon}{2})} \in \mathcal{L}(H), \end{cases}$

(4.3.2b)

with [ ]′ denoting duality with respect to H as a pivot space. Thus, (4.3.2b) and analyticity of $e^{\mathcal{A}t}$ in turn imply, via (4.3.0), that

(4.3.3a) $\quad \begin{cases} \left|Qe^{\mathcal{A}t}\right|_{\mathcal{L}(H)} = \left|e^{\mathcal{A}^*t}Q^*\right|_{\mathcal{L}(H)} \leq \frac{Ce^{-\delta t}}{t^{\frac{3}{4}+\frac{\epsilon}{2}}}, \ t > 0, \\ \left|Qe^{\mathcal{A}t}\right|_{\mathcal{L}(\mathcal{D}((-\mathcal{A})^{\frac{1}{4}+\frac{\epsilon}{2}});H)} = \left|e^{\mathcal{A}^*t}Q^*\right|_{\mathcal{L}(H;[\mathcal{D}((-\mathcal{A})^{\frac{1}{4}+\frac{\epsilon}{2}})]')} \\ \qquad\qquad\qquad \leq \frac{Ce^{-\delta t}}{t^{\frac{1}{2}-\frac{\epsilon}{4}}}, \ t > 0, \end{cases}$

(4.3.3b)

$$\mathcal{D}((-\mathcal{A})^{\frac{1}{4}+\frac{\epsilon}{2}}) = (H_0^{\frac{1}{2}+\epsilon}(\Omega))^d \cap H \subset W \equiv (H^{\frac{1}{2}+\epsilon}(\Omega))^d \cap H,$$

recalling (1.15). Indeed, as for (4.3.3a), we see that it readily follows by writing $Qe^{\mathcal{A}t}x = Q(-\mathcal{A})^{-(\frac{3}{4}+\frac{\epsilon}{2})}(-\mathcal{A})^{(\frac{3}{4}+\frac{\epsilon}{2})}e^{\mathcal{A}t}x$, $x \in H$, for the first estimate on the left, whereby then the second estimate on the right follows by duality. Similarly, as for (4.3.3b), one first takes $x \in \mathcal{D}((-\mathcal{A})^{\frac{1}{4}+\frac{\epsilon}{2}}) \equiv (H_0^{\frac{1}{2}+\epsilon}(\Omega))^d \cap H$ by (1.15), that is, $y = (-\mathcal{A})^{\frac{1}{4}+\frac{\epsilon}{2}}x \in H$, and likewise rewrites $Qe^{\mathcal{A}t}x = Q(-\mathcal{A})^{-(\frac{3}{4}+\frac{\epsilon}{4})}(-\mathcal{A})^{\frac{1}{2}-\frac{\epsilon}{4}}e^{\mathcal{A}t}y$, whereby then one obtains

(4.3.3c) $\quad \left|Qe^{\mathcal{A}t}x\right|_H \leq C\left|(-\mathcal{A})^{\frac{1}{2}-\frac{\epsilon}{4}}e^{\mathcal{A}t}y\right|_H \leq \frac{Ce^{-\delta t}}{t^{\frac{1}{2}-\frac{\epsilon}{4}}}|y|_H = \frac{Ce^{-\delta t}}{t^{\frac{1}{2}-\frac{\epsilon}{4}}}|x|_{\mathcal{D}((-\mathcal{A})^{\frac{1}{4}+\frac{\epsilon}{2}})},$

and the estimate on the left side of (4.3.3b) is proved, from which the estimate on the right side of (4.3.3b) follows by duality. Since $H_0^{\frac{1}{2}+\epsilon}(\Omega)$ is strictly contained in $H^{\frac{1}{2}+\epsilon}(\Omega)$ [**L-M.1**, p. 55], we are *not* authorized to extend the above estimate to all of $W \equiv (H^{\frac{1}{2}+\epsilon}(\Omega))^d \cap H$.

REMARK 4.3.1. Property (i) above for $QS(t)x \in L^2(0,\infty;H)$, $x \in W$, combined with (4.3.3b), yields that $R_1 \in \mathcal{L}(W;[(H_0^{\frac{1}{2}+\epsilon}(\Omega))^d \cap H]')$. This is as much as one can extract from a *direct* analysis of formula (4.3.1) for R_1, for $x \in H$. Alternatively, Lemma 4.3.1(i) below will show that $R_1 \in \mathcal{L}(\mathcal{D}(A_R);H)$. From established Optimal Control Theory [**F-L-T.1**], [**L-T.1**], [**L-T.2**], one expects, however, that: (a) the operator R_1 will have to coincide with the operator R defined in (4.1.8); and hence, accordingly, (b) that $R_1 \in \mathcal{L}(W;W')$, a serious refinement over the result $R_1 \in \mathcal{L}(W;[(H_0^{\frac{1}{2}+\epsilon}(\Omega))^d \cap H]')$, obtained above by direct computations on (4.3.1). In fact, both expectations (a) and (b) will turn out to be true. These will be established in Section 4.4, however, by first identifying R_1 with R via the optimal dynamics, and hence concluding that $R_1 \in \mathcal{L}(W;W')$, using the optimization theory, not the direct analysis of formula (4.3.1). All this shows the challenge of the present OCP with high observation at the $(H^{\frac{3}{2}+\epsilon}(\Omega))^d$-topological level.

Further properties of the operator R_1, motivated by established theory on OCP [**F-L-T.1**], [**L-T.1**], [**L-T.2**], are given below.

LEMMA 4.3.1. *With reference to R_1 in (4.3.1), we have*

(i) $(-\mathcal{A}^*)^{\frac{1}{4}-\epsilon}R_1$: continuous $\mathcal{D}(A_R) \to H$, $\epsilon > 0$; namely

(4.3.4) $|R_1 x|_{(H^{\frac{1}{2}-2\epsilon}(\Omega))^d \cap H} = |(-\mathcal{A}^*)^{\frac{1}{4}-\epsilon}R_1 x|_H \leq c|A_R x|_W; \ \forall \ x \in \mathcal{D}(A_R)$

$$\text{or } (-\mathcal{A}^*)^{\frac{1}{4}-\epsilon}R_1 A_R^{-1} \in \mathcal{L}(W;H).$$

(ii) for x as indicated, we have respectively

(4.3.5a) $\quad \mathcal{A}^* R_1 x = -Q^* Q x - R_1 A_R x \in [(H^{\frac{3}{2}+\epsilon}(\Omega))^d \cap H]', \text{ if } x \in \mathcal{D}(A_R^2);$

(the identity is actually true just for $x \in \mathcal{D}(A_R)$)

(4.3.5b) $(-\mathcal{A}^*)^{\frac{1}{4}-\epsilon}R_1 A_R x \in H, \ |(-\mathcal{A}^*)^{\frac{1}{4}-\epsilon}R_1 A_R x|_H \leq c|A_R^2 x|_W, \text{ if } x \in \mathcal{D}(A_R^2);$

(4.3.5c) $\quad\quad\quad Q^* Q x \in [(H^{\frac{3}{2}+\epsilon}(\Omega))^d \cap H]' \subset [\mathcal{D}(A_R)]', \text{ if } x \in \mathcal{D}(A_R);$

where $\mathcal{D}(A_R) \subset (H^{\frac{3}{2}+\epsilon}(\Omega))^d \cap H \subset W$, by (4.2.10).

(iii) For $x \in \mathcal{D}(A_R^2)$ or $A_R x \in \mathcal{D}(A_R)$, we have (duality $[\ ]'$ with respect to H as a pivot space)

(4.3.6) $\quad\quad |\mathcal{A}^* R_1 x|_{[H^{\frac{3}{2}+\epsilon}(\Omega) \cap H]'} \leq C|A_R^2 x|_W, \quad \forall \ x \in \mathcal{D}(A_R^2);$

PROOF. (i) Estimate (4.3.4) follows at once from the definition (4.3.1) by simply invoking property (4.2.11) for $QS(t)x$, $x \in \mathcal{D}(A_R)$, as well as (4.3.2b) and analyticity of $e^{\mathcal{A}^*t}$. We thus obtain for $x \in \mathcal{D}(A_R)$:

(4.3.7) $|(-\mathcal{A}^*)^{\frac{1}{4}-\epsilon}R_1 x|_H$

$$\leq \int_0^\infty \left|(-\mathcal{A}^*)^{1-\frac{\epsilon}{2}} e^{\mathcal{A}^* t}\right|_{\mathcal{L}(H)} |(-\mathcal{A}^*)^{-\frac{3}{4}-\frac{\epsilon}{2}} Q^*|_{\mathcal{L}(H)} |QS(t)x|_H dt$$

$$\leq c \int_0^\infty \frac{e^{-\delta t}}{t^{1-\frac{\epsilon}{2}}} e^{-at} |A_R x|_W dt \leq C|A_R x|_W, \ x \in \mathcal{D}(A_R),$$

and (4.3.4) is established.

(ii) Following [**F-L-T.1**], we have that the additional regularity of the elements $x \in \mathcal{D}(A_R)$ allows for integration by parts in formula (4.3.1) defining R_1, via (4.2.9):

(4.3.8a) $\quad \mathcal{A}^* R_1 x = \int_0^\infty \mathcal{A}^* e^{\mathcal{A}^* t} Q^* QS(t) x \, dt = \left[e^{\mathcal{A}^* t} Q^* QS(t)x\right]_{t=0}^{t=\infty}$

$$- \int_0^\infty e^{\mathcal{A}^* t} Q^* QS(t) A_R x \, dt, \quad x \in \mathcal{D}(A_R).$$

(4.3.8b) $\quad\quad = -Q^* Q x - R_1 A_R x, \quad x \in \mathcal{D}(A_R).$

In the last step, we have recalled (4.3.0), (4.3.3b), and (4.2.11), as well as (4.3.1) again. Thus, for $x \in \mathcal{D}(A_R)$, we obtain the identity in (4.3.5a) from (4.3.8b), the regularity of whose terms we now present.

To establish (4.3.5c), we first recall (4.3.2a) and (4.2.10), so that the first term on the RHS of (4.3.5) is estimated as follows:

(4.3.9) $\quad\quad |Q^* Q x|_{[(H^{\frac{3}{2}+\epsilon}(\Omega))^d \cap H]'} \leq c|Qx|_H \leq cC|A_R x|_W, \quad x \in \mathcal{D}(A_R).$

As for the term in (4.3.5b), we apply (4.3.4) with $x \in \mathcal{D}(A_R)$ replaced by $A_R x \in \mathcal{D}(A_R)$, to get *a-fortiori*

(4.3.10) $\quad\quad\quad |R_1 A_R x|_H \leq c|A_R^2 x|_W, \quad x \in \mathcal{D}(A_R^2).$

(iii) Finally, using both (4.3.10) and, conservatively, (4.3.9) in (4.3.5a) yields

$$(4.3.11) \quad |\mathcal{A}^* R_1 x|_{[(H^{\frac{3}{2}+\epsilon}(\Omega))^d \cap H]'} \leq |Q^* Q x|_{[(H^{\frac{3}{2}+\epsilon}(\Omega))^d \cap H]'} + |R_1 A_R x|_H$$

$$\leq C|A_R^2 x|_W, \quad x \in \mathcal{D}(A_R^2),$$

and (4.3.11) proves (4.3.6), as desired. □

Next, motivated by available studies of the OCP [**L-T.1**], [**L-T.2**], [**F-L-T.1**], with reference to (3.1.72), (3.1.73) for L_t^* and $\mathcal{K}^*$ and to (4.2.3), we introduce the operator

$$(4.3.12) \quad -L_t^* Q^* Q y^*(\,\cdot\,,0;x) \equiv -L_t^* Q^* QS(\,\cdot\,)x \equiv D^* \mathcal{A}^* \{\mathcal{K}^* Q^* QS(\,\cdot\,)x\}(t)$$

$$\equiv D^* \mathcal{A}^* \int_t^\infty e^{\mathcal{A}^*(\tau-t)} Q^* QS(\tau) x \, d\tau, \quad x \in W, \ t \geq 0$$

$$\equiv D^* \mathcal{A}^* \int_t^\infty e^{\mathcal{A}^*(\tau-t)} Q^* QS(\tau-t) S(t) x \, d\tau$$

(by (4.3.1)) $\equiv D^* \mathcal{A}^* R_1 S(t) x,$

invoking (4.3.1) in the last step. From the available aforementioned literature, we expect the linear operator $L_t^* Q^* Q$ (with respect to $x \in W$) to provide eventually the sought-after feedback synthesis of the corresponding optimal control $u^*(\,\cdot\,,0;x)$. We begin with a regularity property of such operator.

PROPOSITION 4.3.2. *With reference to (4.3.12), we have:*

$$(4.3.13\text{a}) \quad \begin{cases} L_t^* Q^* QS(\,\cdot\,) : \text{continuous } \mathcal{D}(A_R^2) \to (H^{-1-\epsilon}(\Gamma))^d, \text{ with uniform bound in } t \geq 0; \text{ that is, there exists a constant } C > 0, \text{ such that:} \\ \\ (4.3.13\text{b}) \quad |L_t^* Q^* QS(\,\cdot\,)x|_{(H^{-1-\epsilon}(\Gamma))^d} \leq C|A_R^2 x|_W, \quad x \in \mathcal{D}(A_R^2), \ t \geq 0. \end{cases}$$

PROOF. (i) With reference to (4.3.12), the critical step consists in showing that: there exists a constant $c > 0$, such that for all $x \in \mathcal{D}(A_R^2)$, we have

$$(4.3.14) \quad \left| \mathcal{A}^* \int_t^\infty e^{\mathcal{A}^*(\tau-t)} Q^* QS(\tau) x \, d\tau \right|_{[(H^{\frac{3}{2}+\epsilon}(\Omega))^d \cap H]'}$$

$$\leq c|A_R^2 x|_W, \quad x \in \mathcal{D}(A_R^2), \ t \geq 0.$$

To establish (4.3.14), we begin by integrating by parts with $x \in \mathcal{D}(A_R^2)$ (eventually) and thus obtain the counterpart of (4.3.8), again using (4.3.0), (4.2.11):

$$(4.3.15) \quad \mathcal{A}^* \int_t^\infty e^{\mathcal{A}^*(\tau-t)} Q^* QS(\tau) x \, d\tau$$

$$= -Q^* QS(t) x - \int_t^\infty e^{\mathcal{A}^*(\tau-t)} Q^* QS(\tau) A_R x \, d\tau,$$

where, by recalling first (4.3.9) (left) and next (4.2.11), we have for: $x \in \mathcal{D}(A_R)$:

$$(4.3.16) \quad |Q^* QS(t)x|_{[(H^{\frac{3}{2}+\epsilon}(\Omega))^d \cap H]'} \leq C|QS(t)x|_H$$

$$\leq Ce^{-at}|A_R x|_W, \quad x \in \mathcal{D}(A_R), \ t \geq 0.$$

Moreover, by invoking first (4.3.3a) and next again (4.2.11) [or the last step in (4.3.16)], we obtain for all $x \in \mathcal{D}(A_R^2)$,

$$(4.3.17) \quad \left| \int_t^\infty e^{\mathcal{A}^*(\tau-t)} Q^* Q S(\tau) A_R x \, d\tau \right|_H$$

$$\leq \int_t^\infty \left| e^{\mathcal{A}^*(\tau-t)} Q^* \right|_{\mathcal{L}(H)} |QS(\tau) A_R x|_H \, d\tau$$

$$\leq \int_t^\infty \frac{c e^{-\delta(\tau-t)}}{(\tau-t)^{\frac{3}{4}+\frac{\epsilon}{2}}} C e^{-a\tau} |A_R^2 x|_W \, d\tau$$

$$(4.3.18) \quad \leq C |A_R^2 x|_W, \quad x \in \mathcal{D}(A_R^2), \; t \geq 0.$$

Finally, using estimates (4.3.16) and (4.3.18) in (4.3.15), we readily obtain estimate (4.3.14), as desired.

(ii) We next invoke (3.1.3d) on the regularity of D, which we use with $s = 1+\epsilon$: thus taking $u \in (H^{1+\epsilon}(\Gamma))^d$, $u \cdot \nu \equiv 0$ as required by the definition of U in (4.1.1a), and $x \in \mathcal{D}(A_R^2)$, we estimate recalling (3.1.28) on L_t^*:

$$(4.3.19) \quad \left| (L_t^* Q^* Q S(\,\cdot\,) x, u)_{(L^2(\Gamma))^d} \right|$$

$$(\text{by } (3.1.28)) \;=\; \left| \left(D^* \mathcal{A}^* \int_t^\infty e^{\mathcal{A}^*(\tau-t)} Q^* Q S(\tau) x \, d\tau, u \right)_{(L^2(\Gamma))^d} \right|$$

$$=\; \left| \left(\mathcal{A}^* \int_t^\infty e^{\mathcal{A}^*(\tau-t)} Q^* Q S(\tau) x \, d\tau, Du \right)_H \right|$$

$$(\text{by } (4.3.14)) \;\leq\; C |A_R^2 x|_W |Du|_{[H^{\frac{3}{2}+\epsilon}(\Omega) \cap H]}$$

$$(4.3.20) \quad (\text{by } (3.1.3d)) \;\leq\; C |A_R^2 x|_W |u|_{(H^{1+\epsilon}(\Gamma))^d}, \quad x \in \mathcal{D}(A_R^2), \text{ uniformly in } t \geq 0,$$

$u \cdot \nu \equiv 0$ on Σ. Thus, by duality, estimate (4.3.20) precisely yields estimate (4.3.13), as desired. □

We can now provide the sought-after feedback synthesis of the corresponding optimal control $u^*(\,\cdot\,, 0; x)$, initially for $x \in \mathcal{D}(A_R^2)$.

PROPOSITION 4.3.3. *Let $u^*(t, 0; x)$, $x \in W$, so that $u^* \cdot \nu = 0$ on Γ, be the optimal control of the OCP (4.1.2)–(4.1.4), guaranteed by Corollary 4.1.2, with regularity property given by (4.1.10). Let now $x \in \mathcal{D}(A_R^2)$, $v \in (H^{1+\epsilon}(\Gamma))^d$, $v \cdot \nu \equiv 0$ on Σ. Then, recalling R_1 in (4.3.1), the following holds true for all $t \geq 0$:*

(i)

$$(4.3.21a) \quad (u^*(t, 0; x), v)_{(L^2(\Gamma))^d} \;=\; -(L_t^* Q^* Q S(\,\cdot\,) x, v)_{(L^2(\Gamma))^d}$$

$$(\text{by } (4.3.12)) \;=\; (D^* \mathcal{A}^* R_1 S(t) x, v)_{(L^2(\Gamma))^d}, \quad x \in \mathcal{D}(A_R^2),$$

$$v \in (H^{1+\epsilon}(\Gamma))^d,$$

that is [recall also $u^*(t,0;x) \in C([0,T];(H^\epsilon(\Gamma))^d)$, $x \in W$ by (4.1.10)]

(4.3.21b) $\quad u^*(t,0;x) = -L_t^* Q^* Q S(\,\cdot\,)x$
$$= D^* \mathcal{A}^* R_1 S(t) x \in C\left([0,T];(H^{-1-\epsilon}(\Gamma))^d\right),$$
$$t \geq 0,\ x \in \mathcal{D}(A_R^2);$$

(ii) there exists a constant $c > 0$ such that

(4.3.22) $\quad |(u^*(t,0;x),v)_{(L^2(\Gamma))^d}| = |(D^* \mathcal{A}^* R_1 S(t)x, v)_{(L^2(\Gamma))^d}|$
$$= |(\mathcal{A}^* R_1 S(t)x, Dv)_H|$$

(4.3.23) $\quad$ *(by (4.3.11))* $\quad \leq c|S(t)A_R^2 x|_W |Dv|_{[(H^{\frac{3}{2}+\epsilon}(\Omega))^d \cap H]}$

(4.3.24) $\quad$ *(by (4.2.4))* $\quad \leq C|A_R^2 x|_W |v|_{(H^{1+\epsilon}(\Gamma))^d}$, $x \in \mathcal{D}(A_R^2)$,
$$v \in (H^{1+\epsilon}(\Gamma))^d,\ \textit{uniformly in } t \geq 0.$$

PROOF. (i) The first identity (left) in (4.3.21b) is the classical optimality condition [**L-T.1**] via (4.2.3), adapted to the context of presently available regularity theory. Then, the second identity (on the right) of (4.3.21b) follows at once by (4.3.12).

(ii) The estimates in (4.3.22)–(4.3.24) follow via (4.3.11) [with $x \in \mathcal{D}(A_R^2)$ replaced by $S(t)x \in \mathcal{D}(A_R^2)$ for which (4.2.4) holds true] and (3.1.3d) for D, with $s = 1 + \epsilon$. $\square$

The culmination of the present section is the following 'feedback synthesis' result, valid for all $x \in W$, which extends Proposition 4.3.3.

PROPOSITION 4.3.4. *(i) There exists a (feedback) operator*
$$F \equiv D^* \mathcal{A}^* R_1 \in \mathcal{L}(\mathcal{D}(A_R^2); (H^{1+\epsilon}(\Gamma))^d),$$
$(Fx) \cdot \nu = 0$ *on* Γ *for* $x \in \mathcal{D}(A_R^2)$, *such that the (unique) optimal control* $u^*(t,0;x)$ *is given by*

(4.3.25) $\quad u^*(t,0;x) = Fy^*(t,0;x) = FS(t)x$
$$= y^*(t,0;x)|_\Gamma \in L^2(0,\infty; (H^{1+\epsilon}(\Gamma))^d$$
$$\cap\, C([0,T]; (H^{1+\epsilon}(\Gamma))^d),\ x \in \mathcal{D}(A_R^2),\ t \geq 0.$$

$(FS(t)x) \cdot \nu = 0$ on Γ, $t \geq 0$, $x \in \mathcal{D}(A_R^2)$.

(ii) With such operator F, one can characterize more explicitly the infinitesimal generator A_R of the s.c. (uniformly stable) semigroup $S(t)$ on W [given in (4.2.9)] by

(4.3.26) $\quad A_R x = \mathcal{A} x - \mathcal{A} D F x = \mathcal{A}[x - DFx] \in \mathcal{D}(A_R),\ x \in \mathcal{D}(A_R^2).$

REMARK 4.3.2. We shall see below in Appendix D that the operator $FS(t)$, originally defined $\mathcal{D}(A_R^2) \to L^2(0,\infty; (H^{1+\epsilon}(\Gamma))^d)$ can be extended, as a block, as an operator $W \to L^2(0,\infty; (L^2(\Gamma))^d)$. We prefer, however, to maintain $F \equiv D^* \mathcal{A}^* R_1 \in \mathcal{L}(\mathcal{D}(A_R^2); (H^{1+\epsilon}(\Gamma))^d)$ as long as possible in our development, and certainly in the present Section 4, and defer the extension of $FS(t)$ to a later stage, in Section 5.1. A similar situation, where the feedback operator F, when applied to the

4.3. FEEDBACK SYNTHESIS VIA THE RICCATI OPERATOR

dynamics $S(t)$ admits a dynamic extension, occurs of course in the hyperbolic-like case [**L-T.2**].

PROOF. (i) We first recall from (4.1.10) that in particular,

(4.3.27) $\quad u^*(t, 0; x) = y^*(t, 0; x)|_\Gamma \in C([0, T]; (H^\epsilon(\Gamma))^d), \; x \in W \subset H,$

so that, setting $t = 0$, we obtain

(4.3.28) $\quad u^*(0, 0; x) = y^*(0, 0; x)|_\Gamma = x|_\Gamma \in (H^\epsilon(\Gamma))^d, \quad x \in W, \quad x|_\Gamma \cdot \nu = 0.$

On the other hand, specializing likewise (4.3.21b) for $t = 0$, we introduce an operator F on $\mathcal{D}(A_R^2)$ by setting

(4.3.29) $\quad u^*(0, 0; x) = D^* \mathcal{A}^* R_1 x \equiv F x \in (H^{-1-\epsilon}(\Gamma))^d, \quad x \in \mathcal{D}(A_R^2).$

Thus, combining (4.3.28) and (4.3.29) at least for $x \in \mathcal{D}(A_R^2)$ and recalling that the map $x \in W \equiv (H^{\frac{1}{2}+\epsilon}(\Omega))^d \cap H \to x|_\Gamma \in (H^\epsilon(\Gamma))^d$ is continuous by trace theory, we obtain

(4.3.30) $\quad u^*(0, 0; x) = F x = D^* \mathcal{A}^* R_1 x = x|_\Gamma \in (H^{1+\epsilon}(\Gamma))^d,$

$$x \in \mathcal{D}(A_R^2) \subset \mathcal{D}(A_R) \subset (H^{\frac{3}{2}+\epsilon}(\Omega))^d \cap H,$$

a-fortiori from (4.2.10), and

(4.3.31) $\quad |Fx|_{(H^{1+\epsilon}(\Gamma))^d} \leq C |x|_W, \; \forall \, x \in \mathcal{D}(A_R^2), \; (Fx) \cdot \nu = 0 \text{ on } \Gamma.$

REMARK 4.3.3. Since $\mathcal{D}(A_R^2)$ is dense in W, it follows from (4.3.31) that F admits a unique continuous extension $F_e \in \mathcal{L}(W; U)$, defined by $F_e x = \lim_{n \to \infty} F x_n$, where $\{x_n\}$ is any sequence with $x_n \in \mathcal{D}(A_R^2)$, and $x_n \to x$ in W. *A-fortiori*, by trace theory, we have $x_n|_\Gamma \to x|_\Gamma$. Using this as well as $F x_n = x_n|_\Gamma$ by (4.3.30), it then follows that $F_e x = x|_\Gamma$, $x \in W$. This continuous extension F_e on all of W is useless, however. When applied to the dynamics, it produces the trivial identity $y|_\Gamma = y|_\Gamma$. We shall need to extract a more significant (unbounded) extension of F given by (4.3.30). See (4.4.20) below, as well as Remark 4.3.2.

Eqn. (4.3.30) establishes (4.3.25) for $t = 0$. Next, one can then propagate the feedback relation in (4.3.30) from $t = 0$ to an arbitrary $t > 0$, by use of the optimal dynamics. In fact, invoking first the identity in (4.2.6) (right) with $s = t$, and then (4.3.30) with x replaced by $y^*(t, 0; x) = S(t)x \in C([0, T]; \mathcal{D}(A_R^2))$ for $x \in \mathcal{D}(A_R^2)$, we obtain

(4.3.32) $\quad u^*(t, 0; x) = u^*(0, 0; y^*(t, 0; x)) = F y^*(t, 0; x) F S(t) x$

$$= F S(t) x = y^*(t, 0; x)|_\Gamma, \; x \in \mathcal{D}(A_R^2),$$

$$\in L^2(0, \infty; (H^{1+\epsilon}(\Gamma))^d) \cap C([0, T]; (H^{1+\epsilon}(\Gamma))^d).$$

The first regularity statement follows *a-fortiori* from (4.1.10), while the second regularity statement follows *a-fortiori* from $S(t)x \equiv y^*(t, 0; x) \in C([0, T]; \mathcal{D}(A_R^2))$ for $x \in \mathcal{D}(A_R^2) \subset \mathcal{D}(A_R) \subset (H^{\frac{3}{2}+\epsilon}(\Omega))^d \cap H$ by (4.2.10) and trace theory. Thus, (4.3.25) is established in full.

(ii) Eqn. (4.3.26) is an immediate consequence of (4.3.25) being substituted in (4.1.2) for the optimal pair. Moreover, for $x \in \mathcal{D}(A_R^2) \subset \mathcal{D}(A_R) \subset (H^{\frac{3}{2}+\epsilon}(\Omega))^d \cap H$, via (4.2.10), it follows that $x - DFx = x - D(x|_\Gamma) \in H^{\frac{3}{2}+\epsilon}(\Omega) \cap H$ by (3.1.3) and trace theory and $[x - DFx]_\Gamma = x|_\Gamma - Fx = 0$. □

74   4. BOUNDARY FEEDBACK UNIFORM STABILIZATION OF LINEARIZED SYSTEM

COROLLARY 4.3.5. *With reference to the operator F introduced in Proposition 4.3.4, the following regularity properties hold true.*
(i)

$$|FS(t)x|_{(H^{1+\epsilon}(\Gamma))^d} \leq Me^{-at}|A_R x|_W, \ t \geq 0, \ \forall \ x \in \mathcal{D}(A_R^2). \tag{4.3.33}$$

(ii) *Recalling from (4.2.3) that* $y^*(t,0;x) = S(t)x$, *we have*

$$\begin{aligned}
(4.3.34) \quad \left|Q\left[S(t)x - e^{\mathcal{A}t}x\right]\right|_H &+ \left|Q\mathcal{A}\int_0^t e^{\mathcal{A}(t-\tau)} Du^*(\tau,0;x)d\tau\right|_H \\
&= \left|Q\left[S(t)x - e^{\mathcal{A}t}x\right]\right|_H + \left|Q\mathcal{A}\int_0^t e^{\mathcal{A}(t-\tau)} DFS(\tau)x \, d\tau\right|_H \\
&\leq C\left\{C_T|A_R x|_W + \frac{e^{-\delta t}}{t^{\frac{1}{2}+\epsilon}}|x|_W\right\}, \quad 0 \leq t \leq T, \ x \in \mathcal{D}(A_R^2);
\end{aligned}$$

PROOF. (i) Estimate (4.3.33) is a restatement of (4.2.12) via $u^*(t,0;x) = FS(t)x$, by (4.3.25) for $x \in \mathcal{D}(A_R^2)$.

(ii) We begin by establishing estimate (4.3.34) for the first term in its LHS. Let $x \in W \equiv (H^{\frac{1}{2}+\epsilon}(\Omega))^d \cap H \subset \mathcal{D}((-\mathcal{A})^{\frac{1}{4}-\frac{\epsilon}{2}})$ by (1.14), (1.15). By using (4.3.2b) on $Q(-\mathcal{A})^{-(\frac{3}{4}+\frac{\epsilon}{2})}$, (4.3.0) and analyticity of the semigroup, as well as (1.14), we estimate:

$$\begin{aligned}
(4.3.35) \quad |Qe^{\mathcal{A}t}x|_H &= \left|Q(-\mathcal{A})^{-(\frac{3}{4}+\frac{\epsilon}{2})}(-\mathcal{A})^{\frac{1}{2}+\epsilon}e^{\mathcal{A}t}(-\mathcal{A})^{\frac{1}{4}-\frac{\epsilon}{2}}x\right|_H \\
(4.3.36) \quad &\leq \frac{Ce^{-\delta t}}{t^{\frac{1}{2}+\epsilon}}\left|(-\mathcal{A})^{\frac{1}{4}-\frac{\epsilon}{2}}x\right|_H = \frac{Ce^{-\delta t}}{t^{\frac{1}{2}+\epsilon}}|x|_{(H^{\frac{1}{2}-\epsilon}(\Omega))^d \cap H} \\
(4.3.37) \quad &\leq \frac{Ce^{-\delta t}}{t^{\frac{1}{2}+\epsilon}}|x|_W, \ t > 0, \ x \in W.
\end{aligned}$$

Thus, recalling now (4.2.11) for $QS(t)x$, $x \in \mathcal{D}(A_R)$, and invoking (4.3.37), we obtain

$$|Q[S(t)x - e^{\mathcal{A}t}x]|_H \leq C\left[e^{-at}|A_R x|_W + \frac{e^{-\delta t}}{t^{\frac{1}{2}+\epsilon}}|x|_W\right], \ t > 0, \ x \in \mathcal{D}(A_R), \tag{4.3.38}$$

and (4.3.38) establishes *a-fortiori* estimate (4.3.34) for the first term on its LHS.

Next, we now show the estimate in (4.3.34) for the second term on its LHS in the form:

$$\begin{aligned}
(4.3.39) \quad \left|Q\mathcal{A}\int_0^t e^{\mathcal{A}(t-\tau)} DFS(\tau)x \, d\tau\right|_H &\\
&\leq C\left\{\frac{e^{-\delta t}}{t^{\frac{1}{2}+\epsilon}}|x|_W + C_T|A_R x|_W\right\}, \quad 0 \leq t \leq T, \ x \in \mathcal{D}(A_R^2).
\end{aligned}$$

In fact, initially for $x \in \mathcal{D}(A_R^3)$, integrating by parts, we obtain via (4.2.9) and (4.3.25) for $A_R x \in \mathcal{D}(A_R^2)$:

$$(4.3.40) \quad Q\mathcal{A} \int_0^t e^{\mathcal{A}(t-\tau)} DFS(\tau)x \, d\tau = -Q \int_0^t \left(\frac{d}{d\tau} e^{\mathcal{A}(t-\tau)} \right) DFS(\tau)x \, d\tau$$

$$= -QDFS(t)x + Qe^{\mathcal{A}t}DFx$$

$$+ \int_0^t Qe^{\mathcal{A}(t-\tau)} DFS(\tau) A_R x \, d\tau.$$

Regarding the first term on the RHS of (4.3.40), we estimate by virtue of (4.1.1b–c) on Q, (3.1.3d) on D, and (4.3.33) on $FS(t)x$:

$$(4.3.41) \quad |QDFS(t)x|_H = |DFS(t)x|_{[(H^{\frac{3}{2}+\epsilon}(\Gamma))^d \cap H]}$$

$$\leq C|FS(t)x|_{(H^{1+\epsilon}(\Gamma))^d}$$

$$\leq Me^{-at}|A_R x|_W, \quad x \in \mathcal{D}(A_R^2).$$

Next, we pass to the second term on the RHS of (4.3.40), recalling (4.3.2b) on Q, (3.1.3d) on D, and (4.3.31) *a-fortiori* on F. We estimate, via analyticity:

$$(4.3.42) \quad |Qe^{\mathcal{A}t}DFx|_H = |Q(-\mathcal{A})^{-(\frac{3}{4}+\frac{\epsilon}{2})}(-\mathcal{A})^{\frac{1}{2}+\epsilon} e^{\mathcal{A}t} (-\mathcal{A})^{\frac{1}{4}-\frac{\epsilon}{2}} DFx|_H$$

$$(4.3.43) \quad \leq c \left| (-\mathcal{A})^{\frac{1}{2}+\epsilon} e^{\mathcal{A}t} \right|_{\mathcal{L}(H)} |Fx|_{(L^2(\Gamma))^d}, \quad x \in \mathcal{D}(A_R^2)$$

$$(4.3.44) \quad \text{(by (4.3.31))} \leq C \frac{e^{-\delta t}}{t^{\frac{1}{2}+\epsilon}} |Fx|_{(L^2(\Gamma))^d}$$

$$\leq \frac{Ce^{-\delta t}}{t^{\frac{1}{2}+\epsilon}} |x|_W, \quad t > 0, \ x \in \mathcal{D}(A_R^2).$$

Finally, regarding the third term on the RHS of (4.3.40), we invoke estimate (4.3.44) followed by (4.3.33) to get for $x \in \mathcal{D}(A_R^3)$:

$$(4.3.45) \quad \left| \int_0^t Qe^{\mathcal{A}(t-\tau)} DFS(\tau) A_R x \, d\tau \right|_H$$

$$\leq \int_0^t \frac{Ce^{-\delta(t-\tau)}}{(t-\tau)^{\frac{1}{2}+\epsilon}} |FS(\tau)A_R x|_{(L^2(\Gamma))^d} d\tau$$

$$(4.3.46) \quad \text{(by (4.3.33))} \leq \int_0^t \frac{Ce^{-\delta(t-\tau)} Me^{-a\tau}}{(t-\tau)^{\frac{1}{2}+\epsilon}} |A_R x|_W d\tau$$

$$(4.3.47) \quad \leq C_T |A_R x|_W, \ 0 \leq t \leq T, \ x \in \mathcal{D}(A_R^3).$$

Thus, (4.3.41), (4.3.44), and (4.3.47) used in (4.3.40) yield estimate (4.3.39) at first for $x \in \mathcal{D}(A_R^3)$, and next, by extension, for all $x \in \mathcal{D}(A_R^2)$, as desired. Thus, (4.3.39) and (4.3.38) prove estimate (4.3.34), as desired. Corollary 4.3.5 is established. □

COROLLARY 4.3.6. *For each $x \in \mathcal{D}(A_R^2)$, and any $0 < T < \infty$, the following identity holds true*

$$(4.3.48) \qquad QS(t)x = Qe^{\mathcal{A}t}x - Q\mathcal{A} \int_0^t e^{\mathcal{A}(t-\tau)} DFS(\tau)x \, d\tau, \ 0 \leq t \leq T,$$

in the space ${}_\alpha C([0,T];H)$ defined by

$$(4.3.49) \qquad f \in {}_\alpha C([0,T];H) \iff \begin{array}{l} \text{(i)} \ \ f \in C((0,T];H); \\ \text{(ii)} \ \ \sup_{(0,T]} (t^\alpha |f(t)|_H) < \infty, \end{array}$$

where in our present case $\alpha = \frac{1}{2} + \epsilon$.

PROOF. We invoke (4.3.25) for u^* in the variation of parameter formula (3.1.6) corresponding to (4.1.2), (4.3.26), and Corollary 4.3.5. □

4.4. Identification of the Riccati operator R in (4.1.8) with the operator R_1 in (4.3.1)

The next result identifies the Riccati operator $R \in \mathcal{L}(W; W')$ introduced in (4.1.8), self-adjoint on H, with the operator R_1 defined in (4.3.1) and, accordingly, critically improves upon the regularity of R_1 obtained in Remark 4.3.1 only by direct computations on (4.3.1).

PROPOSITION 4.4.1. *The following identification holds true:*
(i)

$$(4.4.1) \qquad R_1 x = Rx \in W', \ \forall \ x \in W; \quad R_1 x = Rx \in H, \ \forall \ x \in \mathcal{D}(A_R);$$

(ii)

$$(4.4.2a) \qquad \mathcal{A}^* R_1 x = \mathcal{A}^* Rx = -Q^* Qx - RA_R x \in \left[\left(H^{\frac{3}{2}+\epsilon}(\Omega)\right)^d \cap H\right]',$$
$$\forall \ x \in \mathcal{D}(A_R).$$

$$(4.4.2b) \qquad R_1 A_R x = R A_R x \in W' = [(H^{\frac{1}{2}+\epsilon}(\Omega))^d \cap H]' \ \text{for} \ x \in \mathcal{D}(A_R);$$

PROOF. (i) Let $x \in \mathcal{D}(A_R)$. We shall first show that

$$(4.4.3) \qquad (Rx, y)_H = (R_1 x, y)_H, \ \forall \ x, y \in \mathcal{D}(A_R^2) \subset W,$$

where we recall that $R_1 : \mathcal{D}(A_R) \to H$ continuously by (4.3.4), after which, since R is continuous (bounded) $W \to W'$, see (4.1.8), we can extend the above identity by density to hold true for all $x, y \in W$. Next, we recall that R is self-adjoint on H, see statement below (4.1.7). Finally, these latter two facts then yield the first identity on the left of (4.4.1) from (4.4.3), while the second identity on the right follows also by virtue of the regularity result in (4.3.4). To complete the proof of part (i), it remains to establish identity (4.4.3). To this end, we invoke preliminarily (4.2.3) and identity (4.3.48) and compute for $x, y \in \mathcal{D}(A_R^2)$:

$$\text{(4.4.4)} \quad \int_0^\infty (Qy^*(t,0;x), Qy^*(t,0;y))_H \, dt$$

$$= \int_0^\infty (QS(t)x, QS(t)y)_H \, dt$$

$$\text{(4.4.5) (by (4.3.48))} = \int_0^\infty (QS(t)x, Qe^{\mathcal{A}t}y)_H \, dt$$

$$- \int_0^\infty \left(QS(t)x, Q\mathcal{A}\int_0^t e^{\mathcal{A}(t-\tau)}DFS(\tau)y \, d\tau\right)_H dt$$

$$\text{(4.4.6)} \quad = \int_0^\infty (e^{\mathcal{A}^*t}Q^*QS(t)x, y)_H \, dt$$

$$- \int_0^\infty \int_0^t \left(QS(t)x, Q\mathcal{A}e^{\mathcal{A}(t-\tau)}DFS(\tau)y\right)_H d\tau \, dt$$

$$\text{(4.4.7)} \quad = \left(\int_0^\infty e^{\mathcal{A}^*t}Q^*QS(t)x \, dt, y\right)_H$$

$$- \int_0^\infty \int_\tau^\infty \left(QS(t)x, Q\mathcal{A}e^{\mathcal{A}(t-\tau)}DFS(\tau)y\right)_H dt \, d\tau$$

$$\text{(4.4.8) (by (4.3.1))} = (R_1 x, y)_H$$

$$- \int_0^\infty \left(D^*\mathcal{A}^*\int_\tau^\infty e^{\mathcal{A}^*(t-\tau)}Q^*QS(t)x \, dt, FS(\tau)y\right)_{(L^2(\Gamma))^d} d\tau$$

$$\text{(4.4.9) (by (3.1.72))} = (R_1 x, y)_H + \int_0^\infty (L_\tau^* Q^* QS(\cdot)x, FS(\tau)y)_{(L^2(\Gamma))^d} d\tau$$

$$\text{(4.4.10)} \quad = (R_1 x, y)_H - \int_0^\infty (u^*(\tau,0;x), u^*(\tau,0;y))_{(L^2(\Gamma))^d} d\tau.$$

In going from (4.4.7) to (4.4.9), we have recalled (4.3.1) on R_1 as well as (3.1.72) on L_τ^*. Moreover, in going from (4.4.9) to (4.4.10) we have recalled both (4.3.21b) for $u^*(\tau, 0; x) = -L_\tau^* Q^* QS(\cdot)x$, $x \in \mathcal{D}(A_R^2)$ as well as (4.3.25) for $u^*(\tau, 0; y) = FS(\tau)y$, $y \in \mathcal{D}(A_R^2)$. Hence, from (4.4.4)–(4.4.10), we obtain

$$\text{(4.4.11)} \quad \int_0^\infty \left[(Qy^*(t,0;x), Qy^*(t,0;y))_H + (u^*(t,0;x), u^*(t,0;y))_{(L^2(\Gamma))^d}\right] dt$$

$$= (R_1 x, y)_H,$$

$x, y \in \mathcal{D}(A_R^2)$. But, for any $x, y \in W$, the LHS of (4.4.11) is precisely equal to $(Rx, y)_H$ by (4.1.8b), (4.1.3). Hence (4.4.3) follows, as desired.

(ii) For $x \in \mathcal{D}(A_R)$, identity (4.3.5a) holds true. In fact, the computation in (4.3.8)—which yields (4.3.5a)—requires only $x \in \mathcal{D}(A_R)$ as noted in (4.3.8b). Moreover, $A_R x \in W$, hence $R_1 A_R x = R A_R x \in W' = [(H^{\frac{1}{2}+\epsilon}(\Omega))^d \cap H]'$, as well as $R_1 x = Rx \in H$, both by part (i). Returning to identity (4.3.5a) with these two relations, we obtain (4.4.2a–b). □

Summary. In light of the identification between R and R_1 given by Proposition 4.4.1, we can rewrite a summary list of relations written initially for the intermediary operator R_1, in terms of the the relevant operator R.

(a)

$$(4.4.12) \quad Rx = \int_0^\infty e^{\mathcal{A}^* t} Q^* Q S(t) x \, dt \in W', \ x \in W \ \text{[from (4.3.1), (4.4.1)]};$$

(b)

$$(4.4.13) \quad |Rx|_{(H^{\frac{1}{2}-2\epsilon}(\Omega))^d \cap H} = |(-\mathcal{A}^*)^{\frac{1}{4}-\epsilon} Rx|_H$$
$$\leq c|A_R x|, \ x \in \mathcal{D}(A_R) \ \text{[from (4.3.4)]};$$

(c) from (4.3.5a), (4.4.2), or (4.3.5c), respectively,

$$(4.4.14) \quad \mathcal{A}^* Rx = -Q^* Q x - R A_R x \in [(H^{\frac{3}{2}+\epsilon}(\Omega))^d \cap H]' \ \text{if } x \in \mathcal{D}(A_R);$$

$$(4.4.15) \quad |\mathcal{A}^* Rx|_{[(H^{\frac{3}{2}+\epsilon}(\Omega))^d \cap H]'} \leq c|A_R x|_W, \ x \in \mathcal{D}(A_R);$$

since now, for $A_R x \in W$, we then have $R A_R x \in W' = [(H^{\frac{1}{2}-\epsilon}(\Omega))^d \cap H]'$ continuously, as established by (4.4.1); and, moreover, $Q^* Q x \in [\mathcal{D}(A_R)]'$, see (4.3.5c), since $(Q^* Q x, y)_H = (Qx, Qy)_H$ is well defined for $x, y \in \mathcal{D}(A_R)$ by (4.2.10). Finally, $W' \subset [\mathcal{D}(A_R)]'$. Thus, (4.4.15b) and (4.4.15b) are proved.

(d)

$$(4.4.16) \quad u^*(0, 0; x) = D^* \mathcal{A}^* Rx = Fx \in (H^{1+\epsilon}(\Gamma))^d, \ x \in \mathcal{D}(A_R^2) \ \text{[from (4.3.30)]}$$

The following result in (4.4.20) below, is important in order to obtain a PDE-interpretation of the abstract linear equation (4.2.9) with generator given by (4.3.26) and, later on, of the abstract non-linear model (5.0) (or (5.1), or (6.0)).

PROPOSITION 4.4.2. *The following identities hold true:*

(i)

$$(4.4.17) \quad Fx \equiv D^* \mathcal{A}^* Rx = -D^* Q^* Q x - D^* R A_R x \in (H^{1+\epsilon}(\Gamma))^d,$$
$$x \in \mathcal{D}(A_R^2), \ (Fx) \cdot \nu = 0 \text{ on } \Gamma;$$

(ii) as to the second term on the RHS of (4.4.17)

$$(4.4.18) \quad D^* R A_R x = D^* \mathcal{A}^* \mathcal{A}^{*-1} R A_R x$$
$$= \nu_0 \frac{\partial}{\partial \nu}(\mathcal{A}^{*-1} R A_R x) \in (L^2(\Gamma))^d, \ x \in \mathcal{D}(A_R^2),$$

where $(D^* R A_R x) \cdot \nu = 0$ on Γ, by Lemma 3.3.1.

(iii) as to the first term on the RHS of (4.4.17)

$$(4.4.19a) \quad D^* Q^* y = D^* \mathcal{A}^* \mathcal{A}^{*-1} Q^* y = \nu_0 \frac{\partial}{\partial \nu}(\mathcal{A}^{*-1} Q^* y) \in (H^{-1-\epsilon}(\Gamma))^d, \ y \in H;$$

$$(4.4.19b) \quad D^* Q^* Q x = \nu_0 \frac{\partial}{\partial \nu}(\mathcal{A}^{*-1} Q^* Q x) \in (H^{-1-\epsilon}(\Gamma))^d, \ x \in (H^{\frac{3}{2}+\epsilon}(\Omega))^d \cap H;$$

(iv)

$$\text{(4.4.20)} \quad Fx = D^*\mathcal{A}^*Rx = \nu_0 \frac{\partial}{\partial \nu}(Rx) \in (H^{1+\epsilon}(\Gamma))^d,$$

$$x \in \mathcal{D}(A_R^2), \ \left(\frac{\partial}{\partial \nu} Rx\right) \cdot \nu = 0 \text{ on } \Gamma.$$

PROOF. (i) Identity (4.4.17) follows from applying the operator D^* across the identity in (4.4.14a) for $x \in \mathcal{D}(A_R^2) \subset (H^{\frac{3}{2}+\epsilon}(\Omega))^d \cap H$ by (4.2.10) and recalling (4.4.16).

(ii) For $x \in \mathcal{D}(A_R^2)$, then $RA_R x \in H$ by (4.4.2b), hence $\mathcal{A}^{*-1}RA_R x \in \mathcal{D}(\mathcal{A}^*)$, and then identity (3.2.2) applies and yields (4.4.18). [Recall $\mathcal{A}^{-1} \in \mathcal{L}(H)$ by (4.3.0)).]

(iii) First, take $z \in (H^{\frac{3}{2}+\epsilon}(\Omega))^d \cap H$, so that $Q^*z \in H$, and then $\mathcal{A}^{*-1}Q^*z \in \mathcal{D}(\mathcal{A}^*)$. Again, applying identity (3.2.2), we obtain

$$\text{(4.4.21)} \quad D^*Q^*z = D^*\mathcal{A}^*\mathcal{A}^{*-1}Q^*z$$

$$= \nu_0 \frac{\partial}{\partial \nu}(\mathcal{A}^{*-1}Q^*z) \in (L^2(\Gamma))^d, \ z \in (H^{\frac{3}{2}+\epsilon}(\Omega))^d \cap H,$$

$(D^*Q^*z) \cdot \nu = 0$ on Γ by Lemma 3.3.1. Next, we prove that

$$\text{(4.4.22)} \quad D^*Q^* : \text{continuous } H \to (H^{-1-\epsilon}(\Gamma))^d.$$

In fact, let $y \in H$, $g \in (H^{1+\epsilon}(\Gamma))^d$, $g \cdot \nu \equiv 0$ on Γ, so that $Dg \in (H^{\frac{3}{2}+\epsilon}(\Omega))^d \cap H$, by (3.1.3d), and $QDg \in H$ by (4.1.1c). Thus,

$$|(D^*Q^*y, g)_{(L^2(\Gamma))^d}| = |(y, QDg)_H| = \text{well-defined}$$

$$\text{(4.4.23)} \quad \leq c|y|_H |g|_{(H^{1+\epsilon}(\Gamma))^d},$$

and (4.4.22) follows. Then, (4.4.22) allows us to extend (4.4.21) as a map with domain all of H and range in $(H^{-1-\epsilon}(\Gamma))^d$ by denseness, so that (4.4.19a) follows and yields (4.4.19b), as $Qx \in H$ for $x \in (H^{\frac{3}{2}+\epsilon}(\Omega))^d \cap H$, by (4.1.1c).

(iv) By (4.17), (4.4.18), (4.4.19b), we obtain for $x \in \mathcal{D}(A_R^2)$, via (4.4.16)

$$\text{(4.4.24)} \quad Fx = D^*\mathcal{A}^*Rx = -\nu_0 \frac{\partial}{\partial \nu}[\mathcal{A}^{*-1}Q^*Qx + \mathcal{A}^{*-1}RA_R x]$$

$$= \nu_0 \frac{\partial}{\partial \nu} Rx \in (H^{1+\epsilon}(\Gamma))^d,$$

where in the last step we have invoked identity (4.4.14a). Then, (4.4.24) and (4.4.16) yield (4.4.20). □

REMARK 4.4.1. In (4.4.20), replace $x \in \mathcal{D}(A_R^2)$ with $S(t)x = y^*(t, 0; x) \in C([0, T]; \mathcal{D}(A_R^2))$. In Appendix C, we establish the following result. Let $x \in W$ and let $x_n \in \mathcal{D}(A_R^2)$, $x_n \to x$ in W. Then, with reference to (4.4.20) we have

$$\text{(4.4.25)} \quad FS(t)x_n = \nu_0 \frac{\partial}{\partial \nu} RS(t)x_n \to FS(t)x = \nu_0 \frac{\partial}{\partial \nu} RS(t)x,$$

in $L^2(0, \infty; (L^2(\Gamma))^d)$, $FS(t)x \cdot \nu = 0$ on Γ. This result will enable us to give a PDE-interpretation of the abstract linear equation (4.2.9) with generator given by (4.3.26) $y_t = \mathcal{A}y - \mathcal{A}Dy$; and, later on, of the abstract non-linear model (5.0).

4.5. A Riccati-type algebraic equation satisfied by the operator R on the domain $\mathcal{D}(A_R^2)$, where A_R is the feedback generator

The culmination of our study of the OCP (4.1.2)–(4.1.4) is the following result, inspired by [**F-L-T.1**], which provides a Riccati algebraic equation, satisfied by the operator R, however, on $\mathcal{D}(A_R^2)$. This will suffice for our purposes—the stabilization result of the forthcoming Section 5.

PROPOSITION 4.5.1. *The following Riccati equation holds true for the operator R in (4.4.12)–(4.4.16), where $U = \{u \in (L^2(\Gamma))^d, u \cdot \nu = 0 \text{ on } \Gamma\}$, as in (4.1.1a):*

(4.5.1a) $\quad (Rx_1, A_R x_2)_H + (RA_R x_1, x_2)_H + (Qx_1, Qx_2)_H = -(Fx_1, Fx_2)_U,$

$$\forall\, x_1, x_2 \in \mathcal{D}(A_R^2),$$

where all terms are well-defined. This is so, since: $Rx_i \in H$ for $x_i \in \mathcal{D}(A_R)$ [by (4.3.4)], R is self-adjoint on H [below (4.1.7)]; $Qx_i \in H$ for $x_i \in \mathcal{D}(A_R)$ [by (4.2.10)]; and, finally, $F \in \mathcal{L}(\mathcal{D}(A_R^2); U)$, U defined in (4.1.1a) [a-fortiori by Proposition 4.3.4]. As a consequence of (4.5.1a), the feedback generator A_R is dissipative on $\mathcal{D}(A_R^2)$, in the norm of $(\,\cdot\,, R\,\cdot\,)_H$ equivalent to $(\,\cdot\,, \cdot\,)_W$; that is,

(4.5.1b) $\quad 2\, Re(A_R x, Rx)_H = -|Qx|_H^2 - |Fx|_U^2 \leq 0, \quad \forall\, x \in \mathcal{D}(A_R^2),$

from which we deduce: $Re(A_R x, Rx)_H \leq 0, \quad \forall\, x \in \mathcal{D}(A_R)$ by denseness of $\mathcal{D}(A_R^2)$ in $\mathcal{D}(A_R)$.

PROOF. STEP 1. We shall invoke the following regularity results:

(4.5.2a) $\quad x_i \in \mathcal{D}(A_R) \Rightarrow x_i \in \left[\left(H^{\frac{3}{2}+\epsilon}(\Omega)\right)^d \cap H\right] \Rightarrow Qx_i \in H$

$\qquad$ by (4.2.10), (4.1.1b-c)

(4.5.2b) $\quad x_i \in \mathcal{D}(A_R^2) \Rightarrow A_R x_i \in \mathcal{D}(A_R) \Rightarrow RA_R x_1 \in \mathcal{D}((-\mathcal{A}^*)^{\frac{1}{4}-\epsilon})$

$\qquad$ by (4.3.4), (4.4.2b), (4.4.13),

(4.5.2c) $\quad x \in \mathcal{D}(A_R) \Rightarrow \mathcal{A}^* Rx_i \in \left[\left(H^{\frac{3}{2}+\epsilon}(\Omega)\right)^d \cap H\right]'$

$\qquad$ by (4.4.14), (4.4.15)

(4.5.2d) $\quad x_i \in \mathcal{D}(A_R^2) \Rightarrow Fx_i = D^* \mathcal{A}^* Rx_i = x_i|_\Gamma = \nu_0 \dfrac{\partial}{\partial \nu} Rx \in (H^{1+\epsilon}(\Gamma))^d$

$\qquad$ by (4.3.30), (4.4.16), (4.4.20),

Our starting point is identity (4.4.14) for $x = x_1 \in \mathcal{D}(A_R^2)$. Here, we take then the H-inner product with $x_2 \in \mathcal{D}(A_R^2)$, add and subtract $(Fx_1, Fx_2)_U$, $U = \{u \in (L^2(\Gamma))^d, u \cdot \nu \equiv 0\}$, and obtain via (4.5.2a–d)

(4.5.3) $(\mathcal{A}^* Rx_1, x_2)_H + (Qx_1, Qx_2)_H + (RA_R x_1, x_2)_H - (Fx_1, Fx_2)_U$

$\qquad = -(Fx_1, Fx_2)_U, \quad x_1 \in \mathcal{D}(A_R^2),\ x_2 \in \mathcal{D}(A_R^2).$

Notice that the first term in (4.5.3) is well-defined as a duality pairing $[(H^{\frac{3}{2}+\epsilon}(\Omega))^d \cap H]' \times [(H^{\frac{3}{2}+\epsilon}(\Omega))^d \cap H]$ by (4.5.2c), (4.5.2a). Moreover, $Fx_i = x_i|_\Gamma \in (H^{1+\epsilon}(\Gamma))^d \cap$

4.5. A RICCATI-TYPE ALGEBRAIC EQUATION

H by (4.5.2d). The remaining two terms in (4.5.3) are well-defined as regular H-inner products, by (4.5.2a), (4.5.2b). Thus, using (4.5.4), we write for $x_i \in \mathcal{D}(A_R^2)$:

(4.5.4) $\quad (\mathcal{A}^* R x_1, x_2)_H - (F x_1, F x_2)_U$

$\qquad\qquad\qquad\quad = (R x_1, \mathcal{A} x_2)_H - (D^* \mathcal{A}^* R x_1, F x_2)_U$

(4.5.5) $\qquad\qquad\qquad\quad = (R x_1, \mathcal{A} x_2)_H - (R x_1, \mathcal{A} D F x_2)_H$

(4.5.6) $\quad$ (by (4.3.26)) $\; = (R x_1, (\mathcal{A} - \mathcal{A} D F) x_2)_H = (R x_1, A_R x_2)_H,$

recalling (4.3.26) in the last step. Each term above is well-defined as a duality pairing. Substituting (4.5.6) in the LHS of identity (4.5.3) yields identity (4.5.1), for $x_i \in \mathcal{D}(A_R^2)$. $\qquad\square$

CHAPTER 5

Theorem 2.3(i): Well-posedness of the Navier-Stokes equations with Riccati-based boundary feedback control. Case $d = 3$

In this section we shall study the local well-posedness of the original Navier-Stokes problem (2.1) under the action of the Riccati-based boundary feedback control $u = Fy$, see (4.3.25):

(5.0) $$y_t = \mathcal{A}y - \mathcal{A}Du - By = (\mathcal{A} - \mathcal{A}DF)y - By,$$

that is, recalling (4.3.26) on A_R:

(5.1) $$y_t = A_R y - By; \qquad y(0) = y^0 \in W \equiv (H^{\frac{1}{2}+\epsilon}(\Omega))^d \cap H.$$

Accordingly, we shall briefly refer to system (5.1) as the *controlled* Navier-Stokes system. We shall deal with the more demanding case $d = 3$. Qualitatively, our goal is *to prove that Eqn. (5.1) admits a unique global (in time) solution, with values in the state space* $W = (H^{\frac{1}{2}+\epsilon}(\Omega))^d \cap H$, *provided that the initial condition y_0 is sufficiently small*. More precisely, we shall establish the following two main results.

THEOREM 5.1. *Let $d = 3$. Let $y^0 \in W \equiv (H^{\frac{1}{2}+\epsilon}(\Omega))^d \cap H$. There is a positive constant $\delta > 0$ (quantitatively identified below in Corollary 5.5, Eqn. (5.37)) such that, if $|R^{\frac{1}{2}}y^0|_H < \delta$, then the abstract N-S Eqn. (5.1) admits a unique fixed-point, mild solution y satisfying $y \in X_T$ where the space X_T is defined by*

(5.2) $$X_T \equiv \{y \in C([0,T];W) \cap L^2(0,\infty;(H^{\frac{3}{2}+\epsilon}(\Omega))^d \cap H); \; Fy \in L^2(0,\infty;U)\},$$

with additional information given in Corollary 5.5, Eqn. (5.38b) below.

The next result takes a smoother I.C. y^0 and delivers a smoother solution y.

THEOREM 5.2. *Let $d = 3$. Let $0 < T < \infty$. There exists a positive constant $\delta > 0$ (quantitatively identified below in Corollary 5.5, Eqn. (5.37), such that:*

(5.3) $$\text{for every I.C. } y^0 \in \mathcal{D}(A_R) \text{ satisfying } |y_0|_W \leq \delta,$$

the abstract N-S (5.1) admits a unique solution y with the following regularity properties:

(5.4a) $\quad \left\{ \begin{array}{l} y \in C([0,T];[(H^{\frac{3}{2}+\epsilon}(\Omega))^d \cap H]) \cap L^2(0,\infty;(H^{\frac{3}{2}+\epsilon}(\Omega))^d \cap H); \end{array} \right.$

(5.4b) $\quad \phantom{\{} y_t \in C([0,T];(H^{\frac{1}{2}+\epsilon}(\Omega))^d \cap H) \cap L^2(0,\infty;(H^{\frac{3}{2}+\epsilon}(\Omega))^d \cap H);$

(5.4c) $\quad \phantom{\{} Fy \in L^2(0,\infty;U); \; By, A_R y \in C([0,T];(H^{\frac{1}{2}+\epsilon}(\Omega))^d \cap H),$

84   5. THEOREM 2.3(I): WELL-POSEDNESS OF THE NAVIER-STOKES EQUATIONS

i.e., recalling $W \equiv (H^{\frac{1}{2}+\epsilon}(\Omega))^d \cap H$; $Z \equiv (H^{\frac{3}{2}+\epsilon}(\Omega))^d \cap H$ from (4.1.1b),

(5.4d) $\quad |y(t)|_Z^2 + |y_t(t)|_W^2 + |By(t)|_W^2 + |A_R y(t)|_W^2 + \int_0^\infty |y(t)|_Z^2 dt$

$$+ \int_0^\infty |y_t(t)|_Z^2 dt + \int_0^\infty |Fy(t)|_U^2 dt \leq C|y^0|_{\mathcal{D}(A_R)}^2, \; y^0 \in \mathcal{D}(A_R),$$

with constant $C > 0$ independent of $\delta > 0$.

Proof of Theorem 5.1. *Step 1.* The proof of both results shall make use of the following regularity result for the corresponding linear part of (5.1).

LEMMA 5.3. *Consider the abstract evolution*

(5.5) $\qquad\qquad z_t = A_R z + f, \qquad z(0) = z_0 \in W.$

Assume

(5.6) $\qquad\qquad z_0 \in W; \quad f \in L^1(0,T;W); \; 0 < T < \infty$ *arbitrary.*

Then, problem (5.5), (5.6) admits a unique mild solution $z(t) = z(t; 0; z_0; f)$ satisfying the following estimates:

(5.7) $\quad \left| R^{\frac{1}{2}} z(t) \right|_H^2 + \int_0^t |Qz(\tau)|_H^2 d\tau + \int_0^t |Fz(\tau)|_U^2 d\tau$

$$\leq \left| R^{\frac{1}{2}} z_0 \right|_H^2 + 2 \int_0^t \left| R^{\frac{1}{2}} f(\tau) \right|_H \left| R^{\frac{1}{2}} z(\tau) \right|_H d\tau;$$

(5.8) $\quad \left| R^{\frac{1}{2}} z \right|_{C([0,T_1];H)}^2 + 2 \int_0^{T_1} |Qz(\tau)|_H^2 d\tau + 2 \int_0^{T_1} |Fz(\tau)|_U^2 d\tau$

$$\leq 2 \left| R^{\frac{1}{2}} z_0 \right|_H^2 + 4 \left[\int_0^{T_1} \left| R^{\frac{1}{2}} f(\tau) \right|_H d\tau \right]^2$$

or recalling the equivalence $|R^{\frac{1}{2}} z|_H \sim |z|_W$ by (4.1.13b):

(5.9) $\quad c|z|_{C([0,T_1];W)}^2 + 2 \int_0^{T_1} |Qz(\tau)|_H^2 d\tau + 2 \int_0^{T_1} |Fz(\tau)|_U^2 d\tau$

$$\leq C \left\{ 2|z_0|_W^2 + 4 \left[\int_0^{T_1} |f(\tau)|_W d\tau \right]^2 \right\},$$

for any $0 < T_1 \leq T$, where c, C are constants related to the norm-equivalence $c|z|_W \leq |R^{\frac{1}{2}} z| \leq C|z|_W$ in (4.1.13b), where the meaning of $Fz(t)$ is given by Appendix D.

PROOF OF LEMMA 5.3. The proof of this lemma employs the Algebraic Riccati Equation (4.5.1) of Proposition 4.5.1.

(i) We assume, at first, smooth data

(5.10) $\qquad\qquad z_0 \in \mathcal{D}(A_R^2), \; f \in L^1(0,T; \mathcal{D}(A_R^2)),$

dense, respectively, in the data assumed in (5.6). We shall first establish the conclusions of Lemma 5.2 for such smooth, dense data as in (5.10), and then extend the results by denseness to the data in (5.6). The data in (5.10) readily imply a unique solution z of (5.5) satisfying

(5.11a) $$\begin{cases} z \in C([0,T]; \mathcal{D}(A_R^2)), \quad z_t \in C([0,T]; W); \\ Rz \in C([0,T]; H). \end{cases}$$
(5.11b)

The regularity in (5.11a) follows by standard semigroup properties on the data in (5.10) via the variation of parameter formula for (5.5). Then, (5.11b) is a consequence of (5.11a) for z, by use of (4.3.4) on $R_1 = R$ (see (4.4.1) (right)) or (4.4.13). Accordingly, the following identity

(5.12) $$(z_t(t), Rz(t))_H - (A_R z(t), Rz(t))_H = (f(t), Rz(t))_H$$

holds true on the real H for such a solution—which is obtained by taking the H-inner product of (5.5) with $Rz(t)$. Since $z(t) \in \mathcal{D}(A_R^2)$ at each t, we may invoke the ARE (4.5.1) with $x_1 = x_2 = z(t)$, and obtain, since R is self-adjoint on H (see below (4.1.7)):

(5.13) $$-2(A_R z(t), Rz(t))_H - |Qz(t)|_H^2 = |Fz(t)|_U^2 \geq 0, \quad z(t) \in \mathcal{D}(A_R^2).$$

This is a dissipativity property of A_R on $z(t)$, in the norm of $(\,\cdot\,, R\,\cdot\,)_H$, equivalent to $(\,\cdot\,, \cdot\,)_W$. Accordingly, we rewrite (5.12) more conveniently, adding and subtracting, as

(5.14) $$\frac{d}{dt}\left|R^{\frac{1}{2}}z(t)\right|_H^2 - \left\{2(A_R z(t), Rz(t))_H + |Qz(t)|_H^2\right\} + |Qz(t)|_H^2$$
$$= 2(f(t), Rz(t))_H,$$

and invoking (5.13) in it, we obtain

(5.15) $$\frac{d}{dt}\left|R^{\frac{1}{2}}z(t)\right|_H^2 + |Qz(t)|_H^2 + |Fz(t)|_U^2 \leq 2(f(t), Rz(t))_H$$
$$\leq 2\left|R^{\frac{1}{2}}f(t)\right|_H \left|R^{\frac{1}{2}}z(t)\right|_H.$$

Integrating (5.15) in time over $\int_0^t$ yields (5.7), as desired, at least for data as in (5.10). To obtain (5.8) from (5.7) we proceed as usual: we take $\max_{[0,T_1]}$ on both sides of (5.7), as well as

(5.16) $$2\int_0^{T_1}\left|R^{\frac{1}{2}}f(t)\right|_H \left|R^{\frac{1}{2}}z(t)\right|_H dt$$
$$\leq 2\left|R^{\frac{1}{2}}z\right|_{C([0,T_1];H)} \int_0^{T_1}\left|R^{\frac{1}{2}}f(t)\right|_H dt$$
$$\leq \epsilon\left|R^{\frac{1}{2}}z\right|_{C([0,T_1];H)}^2 + \frac{1}{\epsilon}\left[\int_0^{T_1}\left|R^{\frac{1}{2}}f(t)\right|_H dt\right]^2,$$

on the right-hand side. Specializing to $\epsilon = \frac{1}{2}$ yields (5.8), as desired, for any $0 < T_1 \leq T$, at least for data as in (5.10).

(ii) A standard denseness argument extends the validity of (5.8) to all data in (5.6), at least with regard to the terms $R^{\frac{1}{2}}z$ and Qz. The term Fz is more delicate and is handled in Appendix D, Proposition D.1, as F, being a trace operator, is not closable. It requires the extension of the term $Fz(t)$, originally defined $\mathcal{D}(A_R^2) \to L^2(0,T;U)$, and then extended to $W \to L^2(0,T;U)$. This includes an analogous extension of the terms $FS(t)z_0$ and $F\int_0^t S(t-\tau)f(\tau)d\tau$ from $z_0 \in \mathcal{D}(A_R^2)$ and $f \in L^2(0,T;\mathcal{D}(A_R^2))$ to $z_0 \in W$ and $f \in L^2(0,T;W)$, respectively. Henceforth, the term $Fz(t)$ is to be meant in the sense of the present extension (given by Proposition D.1 of Appendix D). Lemma 5.3 is established. □

Step 2. Continuing with the proof of Theorem 5.1 and with reference to (5.1), we now introduce a map η,

(5.17) $\qquad \eta : v \to z$, where $z_t = A_R z - Bv, \quad z(0) = y^0 \in W$,

with y^0 given in (5.1) and $v \in C([0,T]; (H^{\frac{3}{2}+\epsilon}(\Omega))^d \cap H)$.

Step 3. For $d = 3$, whereby the embedding $H^s(\Omega) \hookrightarrow L^\infty(\Omega)$ is continuous for $s = \frac{3}{2} + \epsilon > \frac{d}{2} = \frac{3}{2}$, the following estimate on the non-linearity of the N-S equation is critical for the present paper (recall $W = (H^{\frac{1}{2}+\epsilon}(\Omega))^d \cap H$ by (4.1.1b)), recalling B in (1.7):

$$|Bv|_W = |Bv|_{[H^{\frac{1}{2}+\epsilon}(\Omega))^d \cap H]}$$
$$\leq c|(v \cdot \nabla)v|_{[(H^{\frac{1}{2}+\epsilon}(\Omega))^d \cap H]}$$

(5.18a) $\qquad \leq C|v|_{[(H^{\frac{3}{2}+\epsilon}(\Omega))^d \cap H]}|\nabla v|_{[(H^{\frac{1}{2}+\epsilon}(\Omega))^d \cap H]}$

(5.18b) $\qquad \leq C|v|^2_{[(H^{\frac{3}{2}+\epsilon}(\Omega))^d \cap H]} = C|Qv|^2_H,$

$\forall\, v \in (H^{\frac{3}{2}+\epsilon}(\Omega))^d \cap H$. We need to justify the passage to (5.18a). This will be done by invoking (deep) results of the literature on multipliers (pointwise multiplication) in pairs of function spaces.

One source is [**M-S.1**, Proposition, p. 124] on multipliers $M(W_p^m \to W_p^\ell)$, with $p = 2$, $m = \frac{3}{2}+\epsilon$, $\ell = \frac{1}{2}+\epsilon$, so that the assumptions of such Proposition: $p = 2 > 1$ and $mp = (\frac{3}{2}+\epsilon)2 = 3 + 2\epsilon > n = 3$ (dimension) are satisfied. We apply such Proposition with $v \in W_2^{\frac{3}{2}+\epsilon}$ and $\gamma = \frac{\partial v}{\partial x_i} = \partial_i v \in W_2^{\frac{1}{2}+\epsilon}$, to obtain by the definition on the norm of γ in [**M-S.1**, p. 26]:

$$|(\partial_i v)v|_{W_2^{\frac{1}{2}+\epsilon}} = |\gamma v|_{W_2^{\frac{1}{2}+\epsilon}} \leq \|\gamma\|_{M(W_2^{\frac{3}{2}+\epsilon} \to W_2^{\frac{1}{2}+\epsilon})} |v|_{W_2^{\frac{3}{2}+\epsilon}},$$

where [**M-S.1**, Proposition, p. 124] gives

$$\|\gamma\|_{M(W_2^{\frac{3}{2}+\epsilon} \to W_2^{\frac{1}{2}+\epsilon})} \sim |\gamma|_{W_{2,\text{unif}}^{\frac{1}{2}+\epsilon}} = |\partial_i v|_{W_{2,\text{unif}}^{\frac{1}{2}+\epsilon}}$$

(see [**M-S.1**, p. 3] for this latter space). Thus, for a bounded domain Ω, combining the two above results, we obtain (5.18a) in our notation $H^{\frac{3}{2}+\epsilon}(\Omega) = W_2^{\frac{3}{2}+\epsilon}(\Omega)$.

Another (more general) source is [**R-S.1**, Theorem 1(i), p. 190] on multipliers $F_{p,q_1}^{s_1} \cdot F_{p,q_2}^{s_2}$. Here we take: $p = 2$, $q_1 = q_2 = 2$ so that $F_{2,2}^s = W_2^s$ ([**R-S.1**, Proposition (iv), (v), p. 14]); moreover, we take $s_1 = \frac{1}{2} + \epsilon$, $s_2 = \frac{3}{2} + \epsilon$, $n = 3$ (dimension). Thus, the assumptions of [**R-S.1**, Theorem 1, p. 190] are satisfied:

$$s_2 > s_1; \quad s_2 = \left(\frac{3}{2}+\epsilon\right) > \frac{n}{p} = \frac{3}{2}.$$

Accordingly, then, the conclusion of this theorem is that:

$$F_{p,q_1}^{s_1} \cdot F_{p,q_2}^{s_2} \mapsto F_{p,q_1}^{s_1} \quad \text{or} \quad W_2^{\frac{1}{2}+\epsilon} \cdot W_2^{\frac{3}{2}+\epsilon} \mapsto W_2^{\frac{1}{2}+\epsilon},$$

which we apply to $v \cdot \nabla v_i$, $v \in W_2^{\frac{3}{2}+\epsilon}$, $\nabla v_i \in W_2^{\frac{1}{2}+\epsilon}$ to obtain again (5.18a).

5. THEOREM 2.3(I): WELL-POSEDNESS OF THE NAVIER-STOKES EQUATIONS

REMARK 5.1. Estimate (5.18b) says that the non-linearity Bv is controlled by the $(H^{\frac{3}{2}+\epsilon}(\Omega))^d$-norm of v. Thus, for $d = 3$, it is estimate (5.18b) on the non-linearity that dictates the level of the observation operator Q in (4.1.1b–c) for the cost functional (4.1.3) of the OCP. Thus, it is the above multiplier result that motivates the choice of Q.

Step 4. Returning to the map η in (5.17), we see by Step 3, Eqn. (5.18b) that

$$(5.19) \qquad v \in C\left([0,T]; H^{\frac{3}{2}+\epsilon}(\Omega) \cap H\right) \Rightarrow Bv \in C([0,T]; W).$$

Thus, using (5.19) on Bv and (5.17) on $z(0) = y_0 \in W$, we can invoke Lemma 5.3, Eqn. (5.9) with $f = -Bv$, and obtain that

$$(5.20) \qquad \eta: \text{ continuous } v \in C\left([0,T]; H^{\frac{3}{2}+\epsilon}(\Omega) \cap H\right)$$
$$\longrightarrow \begin{cases} z \in C([0,T]; W) \cap L^2(0,T; (H^{\frac{3}{2}+\epsilon}(\Omega))^d \cap H), \\ Fz \in L^2(0,T; U), \end{cases}$$

continuously on the data, for z solution of (5.17).

Step 5. Given $0 < T < \infty$ and $\epsilon > 0$, we introduce the space

$$(5.21) \quad X_T \equiv \{f \in C([0,T]; W) \cap L^2(0,\infty; (H^{\frac{3}{2}+\epsilon}(\Omega))^d \cap H); \; Ff \in L^2(0,\infty; U)\}$$

with U in (4.1.1a). Furthermore, given positive numbers r_1, r_2, we define the following set $\mathcal{S}$ (depending on $r_1, r_2, T, \epsilon > 0$) in X_T by setting (recall the norm-equivalence in (4.1.13b))

$$(5.22) \qquad \mathcal{S} \equiv \left\{ v \in X_T : |R^{\frac{1}{2}}v(t)|_H \leq r_1, \; 0 \leq t \leq T; \right.$$
$$\left. \int_0^\infty |Qv(t)|_H^2 dt + \int_0^\infty |Fv(t)|_U^2 dt \leq r_2^2 \right\}.$$

The set $\mathcal{S}$ is closed and convex.

PROPOSITION 5.4. *With reference to (5.17), let $y^0 \in W$, $|R^{\frac{1}{2}}y^0|_H \leq \delta$, for some positive constant δ.*

(i) Let r_1 be a positive number. Choose first $0 < \delta < \frac{r_1}{\sqrt{2}}$, and then a positive number r_2 such that $0 < r_2^2 < \{2 + [4 + 16k(r_1^2 - 2\delta^2)]^{\frac{1}{2}}\}/8k$, where "$k$" is the constant in (5.23) below. Then, the corresponding set $\mathcal{S}$ in (5.22) is invariant under the map η defined by (5.17): $\eta(\mathcal{S}) \subset \mathcal{S}$.

(ii) Restrict now the number r_2 to satisfy $0 < r_2^2 < \frac{1}{16k}$. Then, the map η in (5.17) is a strict contraction on the set $\mathcal{S}$, in the topology of the space X_T in (5.21).

PROOF. (i) Let $v \in \mathcal{S}$ and consider $z = \eta(v)$, where z solves (5.17). We preliminarily recall estimate (5.18b) for Bv, as well as the norm-equivalence (4.1.13b) and (4.1.1c) on Q:

$$(5.23) \qquad |R^{\frac{1}{2}}Bv|_H \leq C|Bv|_W \leq k|Qv|_H^2.$$

Next, we invoke estimate (5.8) of Lemma 5.3 for problem (5.17), i.e., for problem (5.5) with $f = -Bv$: we obtain

$$(5.24) \quad |R^{\frac{1}{2}}z|^2_{C([0,T];H)} + 2\int_0^T |Qz(t)|^2_H dt + 2\int_0^T |Fz(t)|^2_U dt$$

$$\leq 2|R^{\frac{1}{2}}y^0|^2_H + 4\left[\int_0^T |R^{\frac{1}{2}}Bv|_H d\tau\right]^2.$$

Using (5.23) in (5.24), as well as (5.22), we then obtain for $z = \eta(v)$:

$$(5.25) \quad |R^{\frac{1}{2}}z|^2_{C([0,T];H)} + 2\int_0^\infty |Qz(t)|^2_H dt + 2\int_0^\infty |Fz(t)|^2_U dt$$

$$\leq 2|R^{\frac{1}{2}}y^0|^2_H + 4k\left[\int_0^\infty |Qv|^2_H dt\right]^2$$

$$(5.26) \quad \leq 2\delta^2 + 4kr_2^4,$$

by the assumption on y^0. Then, by (5.22) and (5.26), the invariance of $\mathcal{S}$ under η is assured, as soon as we have

$$(5.27) \quad 2\delta^2 + 4kr_2^4 \leq r_1^2 + 2r_2^2, \quad \text{or} \quad 0 < r_2^2 \leq \frac{2 + \sqrt{4 + 16k(r_1^2 - 2\delta^2)}}{8k},$$

which can indeed be achieved for $2\delta^2 < r_1^2$, as assumed.

(ii) With $\mathcal{S}$ invariant under η as in part (i), let $v_i \in \mathcal{S}$ and $z_i = \eta(v_i)$, $i = 1, 2$. Set

$$(5.28) \quad \tilde{z} \equiv z_1 - z_2, \text{ where then } \tilde{z}_t = A_R \tilde{z} - Bv_1 + Bv_2, \ \tilde{z}(0) = 0.$$

Recalling (1.7) for B and its corresponding basic estimate (5.18a), we preliminarily estimate

$$(5.29) \quad |Bv_1 - Bv_2|_W = |P[(v_1 \cdot \nabla)v_1 - (v_2 \cdot \nabla)v_2]|_W$$

$$(5.30) \quad = |P[((v_1 - v_2) \cdot \nabla)v_1 + (v_2 \cdot \nabla)(v_1 - v_2)]|_W$$

$$(5.31) \quad \leq C\left\{|v_1|_{[(H^{\frac{3}{2}+\epsilon}(\Omega))^d \cap H]} + |v_2|_{[(H^{\frac{3}{2}+\epsilon}(\Omega))^d \cap H]}\right\}|v_1 - v_2|_{(H^{\frac{3}{2}+\epsilon}(\Omega))^d \cap H},$$

From here we then obtain

$$(5.32) \quad |R^{\frac{1}{2}}(Bv_1 - Bv_2)|_H \leq k\{|Qv_1|_H + |Qv_2|_H\}|Q(v_1 - v_2)|_H,$$

invoking also (4.1.1c) for Q and (5.23) in the last step. We now invoke estimate (5.8) for problem (5.28), which is problem (5.5) with $f = Bv_2 - Bv_1$, and zero

5. THEOREM 2.3(I): WELL-POSEDNESS OF THE NAVIER-STOKES EQUATIONS

initial condition. We obtain, via (5.32), the counterpart of (5.24):

$$|R^{\frac{1}{2}}\tilde{z}|^2_{C([0,T];H)} + 2\int_0^\infty |Q\tilde{z}(t)|^2_H dt + 2\int_0^\infty |F\tilde{z}(t)|^2_U dt$$

(5.33)
$$\leq 4\left[\int_0^\infty |R^{\frac{1}{2}}(Bv_2 - Bv_1)|_H dt\right]^2$$

(5.34) (by (5.32))
$$\leq 4k\left\{\int_0^\infty [|Qv_1|_H + |Qv_2|_H]|Q(v_1 - v_2)|_H dt\right\}^2$$

(5.35)
$$\leq 8k\left[\left(\int_0^\infty |Qv_1|^2_H dt\right)\right.$$
$$\left. + \left(\int_0^\infty |Qv_2|^2_H dt\right)\right]\left(\int_0^\infty |Q(v_1 - v_2)|^2_H dt\right),$$

by Schwarz inequality in the last step. Thus, recalling the r_2^2-bounds in (5.22) for $v_i \in \mathcal{S}$, we obtain from (5.35):

(5.36) $|R^{\frac{1}{2}}(z_1 - z_2)|^2_{C([0,T];H)} + 2\int_0^\infty |Q(z_1 - z_2)|^2_H dt + 2\int_0^\infty |F(z_1 - z_2)|^2_U dt$

$$\leq 16k\,r_2^2\left(\int_0^\infty |Q(v_1 - v_2)|^2_H dt\right).$$

Choosing $16k\,r_2^2 < 1$ in (5.36), we obtain that the map $v \to z = \eta(v)$ in (5.17) is a strict contraction on $\mathcal{S}$ in the topology of the space X_T in (5.21). □

By virtue of Proposition 5.4(ii) and the contraction mapping theorem, we have obtained

COROLLARY 5.5. *With k the constant in (5.23), let*

(5.37)
$$0 < \delta^2 < \frac{r_1^2}{2}; \quad r_2^2 < \frac{1}{16k}.$$

Let $y^0 \in W$, $|R^{\frac{1}{2}}y^0|_H < \delta$. Then, there exists a unique solution of the abstract N-S Eqn. (5.1) satisfying $y \in X_T$: that is,

(5.38a) $\quad y \in C([0,T];W) \cap L^2(0,\infty;(H^{\frac{3}{2}+\epsilon}(\Omega))^d \cap H); \quad Fy \in L^2(0,\infty;U),$

or (with constant $C > 0$ independent of $\delta > 0$)

(5.38b) $\quad |y(t)|^2_W + \int_0^\infty |y(t)|^2_{[(H^{\frac{3}{2}+\epsilon}(\Omega))^d \cap H]} dt + \int_0^\infty |Fy(t)|^2_U dt \leq C|y^0|^2_W,$

and, moreover, belonging to the set $\mathcal{S}$, so that

(5.38c) $\quad |R^{\frac{1}{2}}y|_{C([0,T];H)} \leq r_1; \quad \int_0^\infty |Qy(t)|^2_H dt + \int_0^\infty |Fy|^2_U dt \leq r_2^2.$

The proof of Theorem 5.1 is complete.

PROOF OF THEOREM 5.2. Under the setting of Corollary 5.5, there exists a unique solution y of the abstract N-S Eqn. (5.1), satisfying (5.38b). We next differentiate (5.1) in time (in the sense of distributions) for such solution $y(t)$, recalling (1.7) for B. After setting $\bar{y}(t) \equiv y_t(t)$ for such solution $y(t)$, we obtain

(5.39a) $\quad\begin{cases} \bar{y}_t = A_R \bar{y} + f; \ f = -P[(\bar{y} \cdot \nabla)y + (y \cdot \nabla)\bar{y}]; \quad \bar{y}(t) \equiv y_t(t) \\ \bar{y}(0) = y_t(0) = A_R y^0 - B y^0 \in W, \text{ for } y^0 \in \mathcal{D}(A_R), \\ \qquad \|\bar{y}(0)\|_W \leq c \|A_R y^0\|_W, \end{cases}$

(5.39b)

where we have taken $y^0 \in \mathcal{D}(A_R)$ as in assumption (5.3), so that $By^0 \in W$ by (5.18b) via $\mathcal{D}(A_R) \subset (H^{\frac{3}{2}+\epsilon}(\Omega))^d \cap H$ by (4.2.10). Thus, $y_t(0) \in W$, as claimed. As usual, via estimate (5.18b), we obtain for f in (5.39a),

(5.40) $\qquad |R^{\frac{1}{2}} f|_H \leq 2k |Q\bar{y}|_H |Qy|_H,$

as in (5.23) or (5.32). We now invoke estimate (5.8) of Lemma 5.2 for problem (5.39) and obtain

(5.41) $\quad |R^{\frac{1}{2}} \bar{y}|^2_{C([0,T];H)} + 2 \int_0^\infty |Q\bar{y}(t)|^2_H dt + 2 \int_0^\infty |F\bar{y}(t)|^2_U dt - 2|R^{\frac{1}{2}} \bar{y}(0)|^2_H$

$$\leq 4 \left[\int_0^\infty |R^{\frac{1}{2}} f(t)|_H dt \right]^2$$

(5.42) $\quad \text{(by (5.40))} \quad \leq \quad 16k^2 \left[\int_0^\infty [|Q\bar{y}|_H |Qy|_H dt \right]^2$

(5.43) $\qquad\qquad\qquad \leq \quad 16k^2 \left[\int_0^\infty |Qy(t)|^2_H dt \right] \left[\int_0^\infty |Q\bar{y}(t)|^2_H dt \right]$

(5.44) $\quad \text{(by (5.38c))} \quad \leq \quad 16k^2 r_2^2 \int_0^\infty |Q\bar{y}(t)|^2_H dt.$

We have made use of the Schwarz inequality to get (5.43) and recalled (5.38c) on the *a-priori* given solution y in (5.4.5). Thus, recalling $\bar{y} = y_t$, we rewrite (5.44) as

(5.45) $\quad |R^{\frac{1}{2}} y_t|^2_{C([0,T];H)} + 2(1 - 8k^2 r_2^2) \int_0^\infty |Qy_t(t)|^2_H dt$

$$+ \int_0^\infty |Fy_t(t)|^2_U dt \leq 2|R^{\frac{1}{2}} y_t(0)|^2_H.$$

Restricting r_2^2 to get $(1 - 8k^2 r_2^2) > 0$, we then obtain from (5.45) that y_t satisfies the regularity properties

(5.46) $\quad y_t \in C([0,T]; (H^{\frac{1}{2}+\epsilon}(\Omega))^d \cap H) \cap L^2(0,\infty; (H^{\frac{3}{2}+\epsilon}(\Omega))^d \cap H);$

$$Fy_t \in L^2(0,\infty; U),$$

continuously in $\bar{y}(0) \in W$ by (5.39b), recalling the norm-equivalence (4.1.13b) and (4.1.1c) on Q. Finally, from the L^2-regularity of y in (5.38a) and y_t in (5.46) we obtain via [**L-M.1**, Vol. 1, p. 19]

(5.47) $\quad y \in C([0,T]; (H^{\frac{3}{2}+\epsilon}(\Omega))^d \cap H)$, hence $By \in C([0,T]; (H^{\frac{1}{2}+\epsilon}(\Omega))^d \cap H,$

by (5.19), continuously in $\bar{y}(0) \in W$ by (5.39b). Finally, (5.46) on y_t, (5.47) on By, and Eqn. (5.1) yield $A_R y \in C([0,T]; (H^{\frac{1}{2}+\epsilon}(\Omega))^d \cap H)$, continuously in $\bar{y}(0) \in W$ by (5.39b), i.e., in $y_0 \in \mathcal{D}(A_R)$. Theorem 5.2 is thus proved via (5.38), (5.46), (5.47). □

CHAPTER 6

Theorem 2.3(ii): Local uniform stability of the Navier-Stokes equations with Riccati-based boundary feedback control

We return to the controlled Navier-Stokes equation (5.1), rewritten here for convenience:

(6.0) $\quad y_t = A_R y - By$, or $y_t = (\mathcal{A} - \mathcal{A}DF)y - By$; $y(0) = y^0 \in W$,

recalling (4.3.26) on A_R followed by (4.3.29) on F. By Theorem 5.1, Corollary 5.5, respectively by Theorem 5.2, this problem is globally well-posed for $y_0 \in W$, respectively $y_0 \in \mathcal{D}(A_R)$, with $|y_0|_W$ suitably small in both cases. Our final main result is:

THEOREM 6.1. *(i) Assume that the I.C. y^0 satisfies: $|y^0|_W \leq \delta$, for a sufficiently small $\delta > 0$ in the sense of Corollary 5.5, so that the abstract N-S Eqn. (6.0) [or (5.1)] admits a unique global fixed-point, mild solution y satisfying the regularity properties (5.38a); that is, estimate (5.38b). Then, there exist constants $M \geq 1$, $\omega > 0$ (independent of $\delta > 0$) such that such solution $y(t)$ satisfies*

(6.1) $\quad |y(t)|_W \leq M e^{-\omega t} |y^0|_W, \quad t \geq 0, \ y^0 \in W \equiv (H^{\frac{1}{2}+\epsilon}(\Omega))^d \cap H.$

(ii) Assume now that the I.C. y^0 satisfies: $y^0 \in \mathcal{D}(A_R)$ ($\subset Z \equiv (H^{\frac{3}{2}+\epsilon}(\Omega))^d \cap H$, by (4.2.10)), $|y^0|_W \leq \delta$, for a sufficiently small $\delta > 0$ in the sense of Corollary 5.5 (Eqn. (5.37), with the further restriction that $0 < r_2^2 < \frac{1}{8k}$), so that the abstract N-S Eqn. (6.0) [or (5.1)] admits a unique global solution y, satisfying the regularity properties (5.4a-b); that is, estimate (5.4d).

Then, there exist constants $M_1 \geq 1$, $\omega_1 > 0$ (independent of $\delta > 0$) such that such solution $y(t)$ satisfies

(6.2) $\quad |y(t)|_Z \leq M_1 e^{-\omega_1 t} |y^0|_Z, \quad t \geq 0, \ y^0 \in \mathcal{D}(A_R), \ Z \equiv (H^{\frac{3}{2}+\epsilon}(\Omega))^d \cap H.$

PROOF. The strategy for part (i) and part (ii) is exactly the same, starting from either estimate (5.38b) of Corollary 5.5 for part (i), or else from estimate (5.4c) of Theorem 5.2 for part (ii).

(iii) To establish the (local) exponential decay (6.1), we shall employ a classical energy-type strategy for autonomous linear or non-linear systems [**Bal.1**, p. 178]. Specifically, introduce the 'energy' of the solution $y(t)$ of (6.0) by setting

(6.3) $\quad E(t) \equiv |y(t)|_W^2$, and seek to show that: $E(T) \leq \rho E(0), \ \rho < 1,$

for some $T > 0$. After (6.3) (right) has been established, it then follows (in both the linear as well as the non-linear semigroup case, and with the same proof [**Bal.1**, p. 178]) that there exist constants $M \geq 1$, $\omega > 0$, such that the sought-after conclusion (6.1) holds true. In present notation, (6.1) is rewritten as: $E(t) \leq$

$Me^{-\omega t}E(0)$, $t \geq 0$. Thus, it remains to establish the inequality in (6.3) (right). To this end, we return to (5.38b) to read off that

(6.4) $\quad |y(t)|_W \leq C|y^0|_W$, $t \geq 0$; or, more generally, $|y(t)|_W \leq |y(\tau)|_W$, $\quad t \geq \tau$

(had the integration in (5.15) been over $[\tau, t]$, rather than $[0, t]$), as Corollary 5.5 holds true under the present setting. By invoking again estimate (5.38b), we obtain *a-fortiori*, in the notation of (6.3), by use of (6.4):

(6.5) $\quad C|y^0|_W^2 = CE(0) \geq \int_0^T |y(t)|^2_{[(H^{\frac{3}{2}+\epsilon}(\Omega))^d \cap H]} dt$

$$\geq \int_0^T |y(t)|^2_{[(H^{\frac{1}{2}+\epsilon}(\Omega))^d \cap H]} dt$$

(6.6) $\quad$ (by (6.3), (4.1.1b)) $= \int_0^T |y(t)|_W^2 dt$

$$\geq \int_0^T |y(T)|_W^2 dt = T|y(T)|_W^2 = TE(T),$$

or (with constant C independent of T and $\delta > 0$)

(6.7) $$E(T) \leq \frac{C}{T} E(0).$$

Choosing T sufficiently large, one obtains (6.3) (right) from (6.7). Thus, the exponential decay (6.1) is established.

(ii) The proof of the decay (6.2) is exactly the same, starting this time from estimate (5.4c) of Theorem 5.2. Specifically, we introduce the new 'energy' of the solution $y(t)$ of (6.0) by setting

(6.8) $\quad\quad \mathcal{E}(t) \equiv |y(t)|_Z^2, \ Z \equiv (H^{\frac{3}{2}+\epsilon}(\Omega))^d \cap H,$

and seek to show that $\mathcal{E}(T) \leq \rho\mathcal{E}(0)$, $\rho < 1$.

To this end, as Theorem 5.2 holds true under the present setting, we return to (5.4c) to read off that

(6.9) $\quad |y(t)|_Z \leq C|y^0|_Z$, $t \geq 0$; or, more generally, $|y(t)|_Z \leq |y(\tau)|_Z$, $t \geq \tau$

(counterpart of (6.4)). By invoking again estimate (5.4c), we obtain *a-fortiori*, in the notation of (6.8), by use of (6.9) (with constant C independent of T and $\delta > 0$):

(6.10) $\quad C|y^0|_Z = C\mathcal{E}(0) \geq \int_0^T |y(t)|_Z^2 dt \geq \int_0^T |y(T)|_Z^2 dt = T|y(T)|_Z^2 = T\mathcal{E}(0),$

and again (6.8) (right) follows from (6.10) by taking T sufficiently large. Thus, the exponential decay (6.2) is established. $\square$

CHAPTER 7

A PDE-interpretation of the abstract results in Sections 5 and 6

In view of the results of Appendix D (pre-announced in Remark 4.4.1, Eqn. (4.4.25)), the following identification holds true: between the abstract feedback operator F and its differential version

$$(7.1) \quad FS(\cdot)x = D^*\mathcal{A}^*RS(\cdot)x = D^*\mathcal{A}^*Ry^*(\cdot,0;x) = \nu_0 \frac{\partial}{\partial \nu} RS(\cdot)x \in L^2(0,\infty;U),$$

continuously in $x \in W$; $U = \{u \in (L^2(\Gamma))^d,\ u \cdot \nu = 0 \text{ on } \Gamma\}$, as in (4.1.1a). Accordingly, then the following PDE-versions of the various abstract equations may be given.

Translated projected linearized Eqn. (2.4) with $u = Fy$; that is, the abstract Eqn. (4.2.9) with generator given by (4.3.26). The abstract version

$$(7.2) \quad y_t = \mathcal{A}_R y = (\mathcal{A} - \mathcal{A}DF)y, \quad \mathcal{B} = -\mathcal{A}D,$$

admits the PDE-version

$$(7.3a) \quad \begin{cases} y_t - \nu_0 \Delta y + (y_e \cdot \nabla)y + (y \cdot \nabla)y_e = \nabla p & \text{in } Q; \\ (7.3b) \quad \nabla \cdot y = 0 & \text{in } Q; \\ (7.3c) \quad y = \nu_0 \frac{\partial}{\partial \nu} Ry & \text{in } \Sigma; \\ (7.3d) \quad y(x,0) = y^0(x) & \text{in } \Omega. \end{cases}$$

Translated projected non-linear Eqn. (5.0). The abstract equation

$$(7.4) \quad y_t = \mathcal{A}_R y - By = (\mathcal{A} - \mathcal{A}DF)y - By$$

studied in Section 5 and Section 6 admits the PDE-version

$$(7.5a) \quad \begin{cases} y_t - \nu_0 \Delta y + (y \cdot \nabla)y + (y_e \cdot \nabla)y + (y \cdot \nabla)y_e = \nabla p & \text{in } Q; \\ (7.5b) \quad \nabla \cdot y = 0 & \text{in } Q; \\ (7.5c) \quad y = \nu_0 \frac{\partial}{\partial \nu} Ry & \text{in } \Sigma; \\ (7.5d) \quad y(x,0) = y^0(x) = y_0(x) - y_e(x) & \text{in } \Omega. \end{cases}$$

Consequently, recalling the translations $y \to y_e + y$, $p \to p_e + p$, $u \to y_e|_\Gamma + u = u$ performed on the original N-S model (1.1) to obtain problem (2.1), we conclude that returning from (2.1) with $u = Fy$ to (1.1), our final well-posedness and local stabilization results of Sections 5 and 6 refer to the system

$$
\begin{cases}
y_t - \nu_0 \Delta y + (y \cdot \nabla) y = f_e + \nabla p & \text{in } Q; \\
\nabla \cdot y = 0 & \text{in } Q; \\
y = \nu_0 \dfrac{\partial}{\partial \nu} R(y - y_e) & \text{in } \Sigma; \\
y(x, \cdot) = y_0(x) & \text{in } \Omega,
\end{cases}
$$

(7.6a)
(7.6b)
(7.6c)
(7.6d)

which is precisely problem (2.8). It is to this system that Theorem 2.3 applies.

APPENDIX A

Technical Material Complementing Section 3.1

In order to both streamline and focus the exposition of the foundational Section 3.1, and make its topics somewhat self-contained, we present in this Appendix A some basic complementary material centered on two technical issues needed there: (i) the definition of the Leray projector P on appropriate Sobolev spaces of negative index, such as they arise in Section 3.1; (ii) the definition of the "Dirichlet map" D in the general case.

Some of the material included here is known, at least in a less general form—in which case we shall provide definite references; some follows more or less readily from known facts, though we cannot pinpoint specific references for the required statements; and, finally, some is new and based on results of the present paper. An example of the latter case is the definition of the Dirichlet map D on $(H^s(\Gamma))^d$, for $-\frac{1}{2} \leq s < \frac{1}{2}$ in Appendix A.2 (Theorem A.2.1(i)), which is obtained as follows: first for $s = -\frac{1}{2}$ by duality via a critical use of the new result (3.2.2) of Proposition 3.2.1; next, by interpolation between this new case $s = -\frac{1}{2}$ and the case $s = \frac{1}{2}$, which is more or less known in the literature, at least in a less general form. We wish to thank one referee for inducing us to include the material of this Appendix A, in order to further clarify Section 3.1.

A.1. Extension of the Leray Projector P Outside the Space $(L^2(\Omega))^d$

The classical definition of the Leray projector $P: (L^2(\Omega))^d$ onto H was recalled in the paragraph below (1.5b), following established literature [**Te.2**], [**C-F.1**, p. 9]. In this paper (and surely in many other contexts as well), we shall need to extend the definition of the Leray projector P beyond the original space $(L^2(\Omega))^d$. In particular, two types of extension are of interest in Section 3.1. They are singled out in the next two results.

First setting. In this first setting, we shall let $y \in C([0,T];H)$ and $y_t \in L^2(0,T;V')$, $V' =$ dual of V with respect to H as a pivot space, where y_t is defined by

$$(A.1.1) \qquad (y_t(t), \varphi)_{V'V} \equiv \lim_{\tau \to 0} \left(\frac{y(t+\tau) - y(t)}{\tau}, \varphi \right)_H, \quad \forall \, \varphi \in V.$$

Here, $(\ ,\ )_H$ is, as usual, the H- (i.e., $(L^2(\Omega))^d$-) inner product, and $(\ ,\ )_{V'V}$ denotes the duality pairing between V and V', which continuously extends $(\ ,\ )_H$.

PROPOSITION A.1.1. *Let $y \in C([0,T];H)$ and $y_t \in L_2(0,T;V')$, with y_t defined in (A.1.1). Then, we can extend the Leray projector P on y_t by duality so that: $Py_t = y_t \in L^2(0,T;V')$.*

PROOF. Let $\varphi \in V$, so that $P\varphi = \varphi$. The following computations naturally extend by duality the definition of P on y_t, so that $Py_t = y_t$:

$$(A.1.2) \qquad (Py_t(t), \varphi)_{V'V} \equiv \lim_{\tau \to 0} \left(P\frac{y(t+\tau) - y(t)}{\tau}, \varphi \right)_H$$

$$= \lim_{\tau} \left(\frac{y(t+\tau) - y(t)}{\tau}, P\varphi \right)_H$$

$$(A.1.3) \qquad = \lim_{\tau} \left(\frac{y(t+\tau) - y(t)}{\tau}, \varphi \right)_H$$

$$= (y_t(t), \varphi)_{V'V}, \quad \forall \varphi \in V,$$

since $P^* = P$ on $(L^2(\Omega))^d$. Then, (A.1.3) says that $Py_t(t) = y_t(t)$ in V', as desired. $\square$

Second setting. Here we shall extend P to act on $(H^{-1}(\Omega))^d$, so that, if $f \in (H^{-1}(\Omega))^d$, then $Pf \in V'$.

Definition A.1.2: Extension of Leray projector to $H^{-1}(\Omega))^d$.

The Leray projector P, originally defined $P : (L^2(\Omega))^d \to H$ can be naturally extended to a linear continuous operator $P : (H^{-1}(\Omega))^d \to V'$ by the following formula:

$$(A.1.4) \qquad P\mu = \mu_-, \quad \forall \mu \in (H^{-1}(\Omega))^d,$$

where $\mu_- \in V'$ is defined by

$$(A.1.5) \qquad \mu_-(\varphi) = \mu(\varphi), \quad \forall \varphi \in V.$$

Thus, viewing μ as a continuous linear functional on $(H_0^1(\Omega))^d$, we have that μ_- is its restriction on the closed subspace V of $(H_0^1(\Omega))^d$. The element $\mu_- \in V'$ is uniquely defined by formula (A.1.5) and

$$(A.1.6) \qquad \|\mu_-\|_{V'} \leq \|\mu\|_{(H^{-1}(\Omega))^d}.$$

Moreover, if $\mu \in (L^2(\Omega))^d$, then by (A.1.5) and density of V in H, we see that $\mu_- \in H$ and taking into account De Rahm theorem [**Te.1**, p. 14] or Weyl decomposition $\mu = P\mu + \nabla p$. We see by (A.1.5) that $\mu_- = P\mu \in H$. Hence, the operator P defined by (A.1.4) is a well-defined, linear continuous extension from $(H^{-1}(\Omega))^d$ to V' of the original operator $P : (L^2(\Omega))^d \to H$.

We note also that since by De Rham theorem

$$(A.1.7) \qquad V^\perp \equiv \{v' \in V'; \; v'(v) = 0, \quad \forall v \in V\} = \nabla(L^2(\Omega))^d,$$

we have by (A.1.4), (A.1.5) the following decomposition formula on V:

$$(A.1.8) \qquad P\mu = \mu + \nabla p, \quad \forall p \in (L^2(\Omega))^d.$$

Since V is a closed subspace of $(H_0^1(\Omega))^d$, the dual space V' is isometrically isomorphic to the quotient space $(H^{-1}(\Omega))^d / \nabla(L^2(\Omega))^d$ [**T-L.1**, p. 135]. Then, according to (A.1.5) and (A.1.7), $P\mu$ can be equivalently defined by: $P\mu = \phi(\mu)$, where ϕ is the canonical mapping of $(H^{-1}(\Omega))^d$ onto $(H^{-1}(\Omega))^d/\nabla(L^2(\Omega))^d$ [**T-L.1**, p. 72]. In particular, we have:

(A.1.9) $$P(\nabla p) = 0, \quad p \in (L^2(\Omega))^d.$$

REMARK A.1.1. We wish to thank Professor G. P. Galdi (University of Pittsburgh) for pointing out to us that one may find in [**G.1**, Theorem III 5.2 or Corollary III 5.1, pp. 169-70] other more general results related to the structure of the space $(W^{-1,q'}(\Omega))^d$ and its relationship with the dual of $\{f \in W_0^{1,q}(\Omega))^d, \text{ div } f = 0\}$. For $q = 2$, these reproduce (A.1.8).

A.2. Definition and Regularity of the Dirichlet Map in the General Case. Abstract Model

We recall for convenience the basic differential operator $\mathbb{A}$ in (3.1.1):

(A.2.1) $$\mathbb{A}y \equiv -\nu_0 \Delta y + (y_e \cdot \nabla)y + (y \cdot \nabla)y_e,$$

say, for $y \in (H^2(\Omega))^d$. Let k be a constant to be specified below. With $\mathbb{A}$ in (A.2.1), consider the following Stokes/Oseen's problem:

(A.2.2a)
(A.2.2b)
(A.2.2c)
$$\begin{cases} (k + \mathbb{A})\psi = \nabla p^* & \text{in } \Omega; \\ \nabla \cdot \psi \equiv 0, & \text{in } \Omega; \\ \psi|_\Gamma = g, & \text{in } \Gamma, \ g \cdot \nu = 0 \text{ on } \Gamma. \end{cases}$$

THEOREM A.2.1. *(i) Let, at first, $g \in (H^s(\Gamma))^d$, $s \geq \frac{1}{2}$, $g \cdot \nu = 0$ on Γ. Then, at least for $k > 0$ sufficiently large, there exists a unique solution*

(A.2.3) $$\psi \equiv \psi_k \in (H^{s+\frac{1}{2}}(\Omega))^d \cap H$$

of problem (A.2.2) and a corresponding $p^* = p_k^* \in H^{s-\frac{1}{2}}(\Omega)$ unique up to an additive constant. In this case, we can define the operator D_k by setting

(A.2.4a) $$\psi_k = D_k g,$$

and obtain then, recalling that $g \cdot \nu = 0$ on Γ:

(A.2.4b) $$D_k : \text{ continuous } (H^s(\Gamma))^d \to (H^{s+\frac{1}{2}}(\Omega))^d \cap H, \ s \geq \frac{1}{2}.$$

(ii) The dynamic linearized problem (3.1.2) admits the following abstract model

(A.2.5) $$y_t - \mathcal{A}y = \mathcal{A}_k D_k g \in [\mathcal{D}(\mathcal{A}^*)]',$$

where $\mathcal{A}$ here denotes actually the extension, by transposition, $\mathcal{A} : H \to [\mathcal{D}(\mathcal{A}^*)]'$, duality with respect to H as a pivot space, of the original operator $\mathcal{A} = -(\nu_0 A + A_0)$ in (1.11) with compact resolvent on H, while $\mathcal{A}_k$ is defined by: $\mathcal{A}_k \equiv (kI - \mathcal{A})$, as a translation of $(-\mathcal{A})$.

(iii) We have

(A.2.6) $$D_k^* \mathcal{A}_k^* \varphi = -\nu_0 \frac{\partial \varphi}{\partial \nu}, \quad \varphi \in \mathcal{D}(\mathcal{A}^*) = \mathcal{D}(\mathcal{A}),$$

so that, while each ingredient $\mathcal{A}_k$ and D_k depends on the translation, their composition $D_k^* \mathcal{A}_k^*$ does not.

PROOF. *Part (i)*. We consider explicitly the least regular case $s = \frac{1}{2}$. By proceeding classically as in the proof of Lemma 3.1.5 of the text (except that now there is not time variable), given $g \in (H^{\frac{1}{2}}(\Gamma))^d$, there exists an interior extension $g_e \in (H^1(\Omega))^d$, by surjectivity of the trace operator. Setting $f \equiv \text{div } g_e$, we have by the divergence theorem:

$$(A.2.7) \quad f \equiv \text{div } g_e \in (L^2(\Omega))^d; \quad \int_\Omega f \, d\Omega = \int_\Gamma g \cdot \nu \, d\Gamma = 0,$$

by (A.2.2c), so that f belongs to the space $L_0^2(\Omega)$ defined in (3.1.34a). As in the proof of Lemma 3.1.5, we next invoke [**Te.1**, Lemma 2.4, p. 32], or (more explicitly) [**K.2**, Lemma 3.2.3, p. 134]. Accordingly, as in (3.1.35a), there exists a unique v:

$$(A.2.8) \quad v \in W^\perp \subset (H_0^1(\Omega))^d, \text{ such that div } v = \text{div } g_e \text{ in } \Omega; \; v|_\Gamma = 0 \; (A.2.8),$$

where $W^\perp$ is the orthogonal complement in $(H_0^1(\Omega))^d$ for the scalar product defined in (3.1.34b) of the space W defined in (3.1.34c). Setting

$$(A.2.9) \quad w = g_e - v \in (H^1(\Omega))^d \cap H, \text{ we get } \nabla \cdot w \equiv 0 \text{ in } \Omega; \; w|_\Gamma = g, \; w|_\Gamma \cdot \nu = 0,$$

recalling (A.2.8). For ψ in (A.2.2) and w in (A.2.9), we introduce the new variable

$$(A.2.10) \quad \varphi = \psi - w.$$

Then, from (A.2.2) for ψ and (A.2.9), (A.2.10), we obtain the problem

$$(A.2.11a) \quad \begin{cases} (k + \mathbb{A})\varphi = \nabla p^* + \rho & \text{in } \Omega; \quad \rho \equiv (k + \mathbb{A})w \in (H^{-1}(\Omega))^d; \\ (A.2.11b) \quad \nabla \cdot \varphi \equiv 0 & \text{in } \Omega; \\ (A.2.11c) \quad \varphi|_\Gamma = 0 & \text{in } \Gamma. \end{cases}$$

We seek a solution φ of problem (A.2.11) for a function $p^* \in L^2(\Omega)$, so that $\nabla p^* \in (H^{-1}(\Omega))^d$. Next, we apply the Leray projector P on (A.2.11a); where we now invoke Definition A.1.2 for the extension of P on $(H^{-1}(\Omega))^d$, in particular, formula (A.1.9). Then, (A.2.11a) becomes

$$(A.2.12) \quad P(k + \mathbb{A})\varphi = P\rho \in V'.$$

Then (A.2.12), along with (A.2.11b-c) so that $P\varphi = \varphi$, can be rewritten abstractly as

$$(A.2.13) \quad (kI - \mathcal{A})\varphi = P\rho \in V' \equiv [\mathcal{D}(A^{\frac{1}{2}})]',$$

after recalling that $\mathcal{A}\varphi = -P\mathbb{A}\varphi$ for the operator $\mathcal{A}$ in (1.11) (and A in (1.6), A_0 in (1.10a), and $\mathbb{A}$ in (A.2.1)). In particular, we have noted below (1.11) that the operator $\mathcal{A}$ is the generator of an analytic C_0-semigroup on H and has compact resolvent with only finitely many eigenvalues with positive or zero real parts. Thus, there are infinitely many values of $k > 0$ such that the operator $(kI - \mathcal{A})$ is boundedly invertible on H, and for each such value of k we obtain from (A.2.13):

$$(A.2.14) \quad \varphi = (kI - \mathcal{A})^{-1}P\rho \in V \equiv \mathcal{D}(A^{\frac{1}{2}}),$$

V defined in (1.3). In (A.2.14) we can take $k = 0$ if $\lambda = 0$ is not an eigenvalue of $\mathcal{A}$; otherwise, if $\lambda = 0$ is an eigenvalue of $\mathcal{A}$, we can take $|k|$ suitably small or else $k > 0$ sufficiently large. Henceforth, we let k be a fixed value for which $(kI - \mathcal{A})^{-1} \in \mathcal{L}(H)$. We then have φ in (A.2.14) depending on k, φ_k, and returning to (A.2.10): $\psi = \varphi + w$, we obtain

$$(A.2.15) \quad \psi \equiv \psi_k = \varphi_k + w \in (H^1(\Omega))^d \cap H,$$

recalling (A.2.14) for $\varphi = \varphi_k$ and (A.2.9) for w. Via (A.2.15), we then define $\psi_k = D_k g$ as in (A.2.4a) and obtain (A.2.4b) in the present case $s = \frac{1}{2}$.

Moreover, by returning to (A.2.12) with ρ as in (A.2.11a), we see that such ψ_k satisfies the equation: (*)

$$P[(k + \mathbb{A})\psi] = 0 \text{ in } V', \text{ as required.}$$

Finally, we need to show that there exists $p^* \in L^2(\Omega)$, such that $(k+\mathbb{A})\psi = \nabla p^*$ in Ω. To this end, we take the duality pairing of the equation in (*) with any $v \in V$, move P across, use that $Pv = v$ and obtain $((k + \mathbb{A})\psi, v)_{V'V} = 0, \forall v \in V$. We can then appeal to De Rham theorem [**Te.1**, Proposition 1.1, p. 14; and also p. 20, top paragraph], [**C-F.1**, Proposition 1.6, p. 7] to conclude that: there exists a scalar distribution p^* such that $(k + \mathbb{A})\psi = \nabla p^*$, with $\nabla p^* \in (H^{-1}(\Omega))^d$. Invoking [**C-F.1**, Proposition 1.7(ii), p. 7], [**Te.1**, Proposition 1.2(ii), p. 14], we conclude that such $p^* \in L^2(\Omega)$, as desired. Part (i) is proved.

Part (ii). We return to the linearized problem (3.1.2), whose first Eqn. (3.1.2a), we rewrite as

(A.2.16) $\qquad y_t + (k + \mathbb{A})y - ky = \nabla p \quad \text{in } Q$.

Subtracting $(k + \mathbb{A})\psi = \nabla p^*$ from both sides of (A.2.16) (recall (A.2.2a)), where $\psi = \psi_k = D_k g$ by (A.2.4a), we obtain, recalling also (3.1.2b,d) and (A.2.2b,c):

(A.2.17a) $\quad \begin{cases} y_t + (k + \mathbb{A})(y - D_k g) - ky = \nabla(p - p^*) & \text{in } Q; \\ \nabla \cdot (y - D_k g) = 0 & \text{in } Q; \\ (y - D_k g)|_\Sigma = g - g = 0 & \text{in } \Sigma. \end{cases}$

(A.2.17b)

(A.2.17c)

Application of the Leray projector P on both sides of (A.2.17a) yields

(A.2.18) $\qquad Py_t + P(k + \mathbb{A})(y - D_k g) - kPy = P\nabla(p - p^*) \text{ in } Q$.

Next, on the LHS of (A.2.18) we recall $Py_t = y_t$ by Proposition A.1.1, $P(ky) = ky$, and recall the operator $\mathcal{A}$ in (1.11) (via (1.6) and (1.10a)); on the RHS of (A.2.18), we invoke Proposition (A.1.2) for the extension of P, more precisely formulas (A.1.10), (A.1.11). We thus obtain by (A.2.18), (A.2.17b–c),

(A.2.19) $\qquad y_t + (k - \mathcal{A})(y - D_k g) - ky = 0$,

or, after a cancellation of the term ky, and taking the *extension*, by transposition, of $\mathcal{A}$ from $H \to [\mathcal{D}(\mathcal{A}^*)]$, duality with respect to H as a pivot space, we obtain

(A.2.20) $\qquad y_t - \mathcal{A}y = (k - \mathcal{A})D_k g \in [\mathcal{D}(\mathcal{A}^*)]'$,

and (A.2.5) is obtained, as desired, with $\mathcal{A}_k = (kI - \mathcal{A})$.

Part (iii). The proof of formula (A.2.6) for $k = 0$ is given in Proposition 3.2.1. For $k \neq 0$, the proof is the same and yields (A.2.6).

Theorem A.2.1 is proved. $\qquad\square$

REMARK A.2.1. To make a connection with known literature, we note that we may take $k = 0$ throughout Theorem A.2.1 in the following cases:
(a) Stokes operator $\mathbb{A}y = -\nu_0 \Delta y$;
(b) Stokes operator perturbed *only* by a first-order perturbation:

$$\mathbb{A}y = -\nu_0 \Delta y + (a \cdot \nabla)y, \quad \nabla \cdot a \equiv 0, \ a \in (H^1(\Omega))^d.$$

For case (a), Theorem A.2.1, Part (i), can be found in the literature [**Te.1**, Proposition 2.3, p. 35, after interpolation of the integer case, which extends Theorem 2.4, p. 31], [**G.1**, Theorem 1.1, p. 188; Theorem 6.1, p. 231]. A corresponding result for the nonlinear Navier-Stokes stationary problem is given in [**F-T.1**, p. 42].

As to case (b), the closest result to our Theorem A.2.1 which we can find is in [**L.1**, Theorem 1.2, p. 191], however, this time with forcing term on the "right-hand side" of the equation, not on the boundary, as in problem (A.2.2). But, for $s \geq \frac{1}{2}$, the boundary datum case can be reduced to the interior ("right-hand side") case, as in our proof of Theorem A.2.1, Part (i).

The above proof of Theorem A.2.1, as well as the considerations above indicate that "generically" one can take $k = 0$. This is, for simplicity of notation, the case chosen in the text of Section 3.1, whereby then (A.2.2), (A.2.4), (A.2.6) reduce to (3.1.3), (3.2.2).

THEOREM A.2.2. *Once Theorem A.2.1(i) defines the 'Dirichlet map' D_k on $(H^s(\Gamma))^d$, $s \geq \frac{1}{2}$, then:*

(i) we can define D_k on $(H^s(\Gamma))^d$, for $s = -\frac{1}{2}$ by duality using formula (A.2.6); and obtain

$$\text{(A.2.21)} \qquad D_k : \text{ continuous } (H^{-\frac{1}{2}}(\Gamma))^d \to (L^2(\Omega))^d;$$

(ii) finally by interpolation between (A.2.21) for $s = -\frac{1}{2}$ and (A.2.4b) for $s = \frac{1}{2}$, we obtain

$$\text{(A.2.22)} \qquad D_k : \text{ continuous } (H^s(\Gamma))^d \to (H^{s+\frac{1}{2}}(\Omega))^d, \quad -\frac{1}{2} \leq s \leq \frac{1}{2};$$

(iii) (A.2.22) combined with (A.2.4b) yields finally

$$\text{(A.2.23)} \qquad D_k : \text{ continuous } (H^s(\Gamma))^d \to (H^{s+\frac{1}{2}}(\Omega))^d, \quad \forall \, s \geq -\frac{1}{2}.$$

PROOF. (i) To define D_k on $(H^{-\frac{1}{2}}(\Omega))^d$, i.e., for $s = -\frac{1}{2}$, we proceed by duality using formula (A.2.6), as follows. Let $f \in (L^2(\Omega))^d$, so that

$$\text{(A.2.24)} \qquad \varphi \equiv (\mathcal{A}_k^*)^{-1} f \in \mathcal{D}(\mathcal{A}^*) \subset (H^2(\Omega))^d,$$

as $\mathcal{A}_k = (k - \mathcal{A})$. Formula (A.2.6) then implies via (A.2.24) and trace theory:

$$\text{(A.2.25)} \qquad D_k^* f = D_k^* \mathcal{A}_k^* (\mathcal{A}_k^*)^{-1} f = D_k^* \mathcal{A}_k^* \varphi = -\nu_0 \left.\frac{\partial \varphi}{\partial \nu}\right|_\Gamma \in (H^{\frac{1}{2}}(\Gamma))^d.$$

Let now $g \in (H^{-\frac{1}{2}}(\Gamma))^d$, so that its duality pairing with $\left.\frac{\partial \varphi}{\partial \nu}\right|_\Gamma \in (H^{\frac{1}{2}}(\Gamma))^d$ is well defined, and then

$$(D_k g, f)_{(L^2(\Omega))^d} = (g, D_k^* f)_{(L^2(\Gamma))^d} = \left(g, -\nu_0 \left.\frac{\partial \varphi}{\partial \nu}\right|_\Gamma\right)_{(L^2(\Gamma))^d}$$

$$\text{(A.2.26)} \qquad\qquad = \text{ well defined}, \, \forall \, f \in (L^2(\Omega))^d.$$

Then, (A.2.26) says that $D_k g \in (L^2(\Omega))^d$, as desired.

Part (i) is proved.

Parts (ii), (iii) are self-explanatory. □

APPENDIX B

Boundary feedback stabilization with arbitrarily small support of the linearized system (3.1.4a) at the $(H^{\frac{3}{2}-\epsilon}(\Omega))^d \cap H$-level, with I.C. $y^0 \in (H^{\frac{1}{2}-\epsilon}(\Omega))^d \cap H$. Cases $d = 2, 3$. Theorem 2.5 for $d = 2$

Orientation. In this Appendix B, we shall consider the problem of achieving *boundary feedback stabilization of the linearized system* (3.1.4a), however, at the topological level of $(H^{\frac{3}{2}-\epsilon}(\Omega))^d \cap H$ in the space variable, with Initial Condition $y^0 \in (H^{\frac{1}{2}-\epsilon}(\Omega))^d \cap H$, $\epsilon > 0$, $d = 2, 3$. As the space norm-level is below the *critical threshold* of $(H^{\frac{3}{2}}(\Omega))^d \cap H$—we can now get the additional benefit that the boundary stabilizing control is allowed to act only on an *arbitrarily small portion* Γ_1 (of positive surface measure) of the boundary $\Gamma = \partial\Omega$, *in full generality*. By contrast, in the case $d = 3$, arbitrarily small boundary support of the stabilizing controller is only claimed at the $(H^{\frac{3}{2}+\epsilon}(\Omega))^d \cap H$-level with I.C. $y^0 \in (H^{\frac{1}{2}+\epsilon}(\Omega))^d \cap H$ in two cases: (i) either under the FDSA (see Theorem 2.2 (based on Theorem 3.7.1)) or else (ii) for I.C. y^0 vanishing on Γ (Theorem 2.1 based on Theorem 3.5.1 and Remark 3.5.1(c).

For $d = 2$, the correct and 'best' topological level is precisely: $y^0 \in (H^{\frac{1}{2}-\epsilon}(\Omega))^d \cap H$ with stabilized solution $y \in (H^{\frac{3}{2}-\epsilon}(\Omega))^d \cap H$ for both the linearized system (3.1.4a) or (3.1.2), as well as the non-linear N-S system (2.1), (2.2), or (1.1).

For $d = 3$, by contrast, an analysis at the $(H^{\frac{3}{2}-\epsilon}(\Omega))^d \cap H$-level is good only for the linearized system (3.1.4a), as an end in itself; while solution of the local stabilization problem for the original N-S system (1.1) requires the higher topological level of $(H^{\frac{3}{2}+\epsilon}(\Omega))^d \cap H$, due to the non-linear term $(y \cdot \nabla)y$ (see (5.18b)): this was repeatedly emphasized above (3.1.22), in Remark 4.1, etc.

B.1. An open-loop infinite dimensional boundary controller $g \in L^2(0, \infty; (L^2(\Gamma_1))^d)$, Γ_1 arbitrary, for the linearized system (3.1.4a) yielding $y(\cdot\,; y_0) \in L^2(0, \infty; (H^{\frac{3}{2}-\epsilon}(\Omega))^d \cap H)$ for $y^0 \in (H^{\frac{1}{2}-\epsilon}(\Omega))^d \cap H$

If we examine the proof of Theorem 3.5.1, it is clear that if we lower the topological level from $[(H^{\frac{3}{2}+\epsilon}(\Omega))^d \cap H]$ to $[(H^{\frac{3}{2}-\epsilon}(\Omega))^d \cap H]$ in the space variable—i.e., below the critical threshold $(H^{\frac{3}{2}}(\Omega))^d \cap H$—then the arguments given there not only simplify but, more importantly, yield a stronger conclusion: that is, that for all I.C. $y^0 \in (H^{\frac{1}{2}-\epsilon}(\Omega))^d \cap H$, the constructed *open-loop* control g can be required to act only on an *arbitrarily small portion* Γ_1 (of positive surface measure) of the boundary $\Gamma = \partial\Omega$. We note that the space $H^{\frac{1}{2}-\epsilon}(\Omega)$ does not recognize boundary conditions. Then the extension from $(H^{\frac{1}{2}-\epsilon}(\Omega))^d \cap H = \mathcal{D}(A^{\frac{1}{4}-\frac{\epsilon}{2}})$ on Ω to $\mathcal{D}(\tilde{A}^{\frac{1}{4}-\frac{\epsilon}{2}})$ on $\tilde{\Omega} = \Omega \cup \omega$ is more direct. We obtain

THEOREM B.1.1. *Let $d = 2, 3$. Let Γ_1 be arbitrary, $\Gamma_1 \subset \Gamma$, $\mathrm{meas}(\Gamma_1) > 0$. There exists an infinite-dimensional open-loop boundary controller g on Γ_1:*

(B.1.1) $$g \in L^2(0, \infty; (L^2(\Gamma_1))^d), \quad g \cdot \nu \equiv 0 \text{ on } \Sigma$$

identified in Eqn. (3.5.27) of Section 3.5—except on Γ_1 rather than Γ, while $g \equiv 0$ on $\Gamma \setminus \Gamma_1$—such that, once inserted in (the Dirichlet B.C. (3.1.2d) of) the linearized problem (3.1.2) or (3.1.4a), yields a solution y with the following properties, for all $\epsilon > 0$:

(B.1.2) $$y^0 \in (H^{\frac{1}{2}-\epsilon}(\Omega))^d \cap H$$
$$\Rightarrow y \in L^2(0, \infty; (H^{\frac{3}{2}-\epsilon}(\Omega))^d \cap H)) \cap H^{\frac{3}{4}-\frac{\epsilon}{2}}(0, \infty; H).$$

B.2. Feedback stabilization in $(H^{\frac{3}{2}-\epsilon}(\Omega))^d$, $d=2,3$, of the linearized system (3.1.4a) with finite-dimensional feedback controller with arbitrary support under FDSA = (3.6.2)

In this section, under the additional FDSA = (3.6.2), we shall obtain a *finite-dimensional* closed-loop feedback stabilizing controller with arbitrarily small support, directly as a Corollary of Section 3.6 and Remark 3.7.1(b). Indeed, in Lemma 3.6.1 and Remark 3.7.1(b), we have seen that the finite-dimensional feedback controller that stabilizes the finite-dimensional projection $z_N(t)$ in (3.4.9) with an arbitrarily large decay rate γ_1 is smooth and given in the form:

(B.2.1) $$v_N(t) = \sum_{i=1}^{K} \left(e^{\bar{A}^u t} P_N z_0, p_i \right)_H w_i \in C^n(0, \infty; \mathcal{F}_1); \quad P_N z_0 = z_N(0),$$

$n = 1, 2, \ldots$, for suitable boundary vectors w_i with arbitrarily small support Γ_1:

(B.2.2) $$w_i \in \mathcal{F}_1 = \mathrm{span}\{\partial_\nu \varphi_{ij}^*|_{\Gamma_1}\}_{i=1, j=1}^{M, \ell_i} \subset (H^{\frac{1}{2}}(\Gamma_1))^d$$

(recall (3.6.9)), which will have to satisfy the rank conditions (3.6.16), so that $w_i \cdot \nu = 0$ on Γ by Lemma 3.3.1, $K = N$ (conservatively) and suitable vectors p_i depending on γ_1. In particular, see (3.6.11),

(B.2.3) $$|v_N(t)|_{(H^{\frac{1}{2}}(\Gamma))^d} + |\dot{v}_N(t)|_{(H^{\frac{1}{2}}(\Gamma))^d} \leq C_{\gamma_1} e^{-\gamma_1 t} |z_0|_{Z_N^u}, \ t \geq 0.$$

Accordingly, such $v_N(t)$ has the additional advantage of forcing an exponential decay onto the infinite-dimensional projection $\zeta_N(t)$ in (3.4.10) with a decay rate arbitrarily close to $|\mathrm{Re}\, \lambda_{N+1}|$, as well as other suitable versions of long-term behavior. These facts, once combined, lead to corresponding feedback stabilization results of the *linearized* (complexified system (3.4.1), hence real) system (3.1.4a) of the original translated model (2.3).

PROPOSITION B.2.1. *Assume the FDSA = (3.6.2) and the rank conditions (3.6.16). For each $y^0 \in H$, there exists a finite-dimensional feedback controller (B.2.1) satisfying (B.2.2), such that the solution z of the corresponding complexified system (3.4.1) on Z, hence real on H, linearized system (3.1.4a) or (3.1.12)*

(B.2.4) $$y_t - \mathcal{A}y = -\mathcal{A}Dg_N, \text{ or } \dot{\eta}_t - \mathcal{A}\eta = -D\dot{g}_N, \ \eta(t) = y(t) - Dg_N(t)$$

B.2. FEEDBACK STABILIZATION IN $(H^{\frac{3}{2}-\epsilon}(\Omega))^d$ OF THE LINEARIZED SYSTEM

$$g_N = Re\ v_N = \sum_{i=1}^{K} Re[(z_N(t), p_i)_H w_i] = \sum_{i=1}^{K} (Re(z_N(t), p_i)_H)(Re\ w_i)$$

$$- \sum_{i=1}^{K} (Im(z_N(t), p_i)_H)(Im\ w_i),$$

satisfies $y \in L^2(0, \infty; H)$ *and in fact:*

(B.2.5) $\quad |A^\theta y(t)| \leq c_{\gamma_0,\delta,\theta} e^{-\gamma_0 t} |y^0|, \quad y^0 \in H,\ \forall\ t \geq \delta > 0,\ 0 \leq \theta < \frac{1}{4},$

where γ_0 is any pre-assigned constant $0 < \gamma_0 < |Re\ \lambda_{N+1}|$, and where we can take $\delta = 0$ for $\theta = 0$.

(ii) Moreover, for $y^0 \in \mathcal{D}(A^{\frac{1}{4}-\epsilon}) = (H^{\frac{1}{2}-2\epsilon}(\Omega))^d \cap H$, such solution satisfies

(B.2.6) $\quad y(\cdot; y_0) \in L^2\left(0, \infty; \mathcal{D}(A^{\frac{3}{4}-\epsilon}) \equiv (H^{\frac{3}{2}-2\epsilon}(\Omega))^d \cap V\right),$

PROOF. (i) To prove (B.2.5) we return to the ζ_N-system (3.4.10) on Z_N^s, thus obtaining

(B.2.7) $\zeta_N(t) = e^{\mathcal{A}_N^s t} \zeta_N(0) + \int_0^t (-\mathcal{A}_N^s)^{\frac{3}{4}+\epsilon} e^{\mathcal{A}_N^s (t-\tau)} (-\mathcal{A}_N^s)^{\frac{1}{4}-\epsilon} (I - P_N) D v_N(\tau) d\tau,$

with v_N given by Lemma 3.6.1, i.e., by (B.2.1), hence with $v_N \cdot \nu \equiv 0$ on Σ. By (3.1.3d),

(B.2.8) $\quad (-\mathcal{A}_N^s)^{\frac{1}{4}-\epsilon}(I - P_N) D \in \mathcal{L}((L^2(\Gamma))^d; H);$

$$|(-\mathcal{A}_N^s)^\theta e^{\mathcal{A}_N^s t}|_{\mathcal{L}(H)} \leq C_{\gamma_0} \frac{e^{-\gamma_0 t}}{t^\theta},\quad 0 < t;\ 0 \leq \theta < 1,$$

for any $0 < \gamma_0 < |Re\ \lambda_{N+1}|$.

Then one proceeds as in [**B-T.1**, Eqn. (3.55)] or [**T.2**] to obtain (B.2.5):

(B.2.9)
$$(-\mathcal{A}_N^s)^\theta \zeta_N(t) = e^{\mathcal{A}_N^s(t-\delta)} (-\mathcal{A}_N^s)^\theta e^{\mathcal{A}_N^s \delta} \zeta_N(0)$$

$$+ \int_0^t (-\mathcal{A}_N^s)^{1-\epsilon'} e^{\mathcal{A}_N^s(t-\tau)} (-\mathcal{A}_N^s)^{\frac{1}{4}-\epsilon} (I - P_N) D v_N(\tau) d\tau,$$

$1 - \epsilon' = \frac{3}{4} + \theta + \epsilon$. Since $|e^{\mathcal{A}_N^s t}|_{\mathcal{L}(H)} \leq C e^{-\gamma_0 t}$, and recalling (B.2.3), we see the integral in (B.2.9) is dominated (brutally) with $\gamma_1 - \gamma_0 > 0$ by

(B.2.10) $\quad C \int_0^t \frac{e^{-\gamma_0(t-\tau)}}{(t-\tau)^{1-\epsilon'}} e^{-\gamma_1 \tau} d\tau \leq C e^{-\gamma_0 t} \int_0^t \frac{e^{-(\gamma_1-\gamma_0)\tau}}{(t-\tau)^{1-\epsilon'}} d\tau$

$$\leq -C e^{-\gamma_0 t} \frac{(t-\tau)^{\epsilon'}}{\epsilon'} \Big|_{\tau=0}^{\tau=t} = \frac{C}{\epsilon'} t^{\epsilon'} e^{-\gamma_0 t},$$

where γ_0 in (B.2.10) may be replaced by $(\gamma_0 + \epsilon'')$ and then (B.2.9) and (B.2.10) yield (B.2.5).

(ii) This time we use model (3.1.12) with $g_N = Re\ v_N$ given by (B.2.1) and essentially mimick the proof of Lemma 3.1.2. Taking the $(I - P_N)$-projection of the η-equation in (B.2.4), we obtain:

$$((I - P_N)\eta)_t = \mathcal{A}_N^s (I - P_N)\eta - (I - P_N) D \dot{g}_N,\text{ or}$$

(B.2.11) $(I - P_N)\eta(t) = (-\mathcal{A}_N^s)^{-(\frac{1}{4}-\epsilon)}e^{\mathcal{A}_N^s t}(-\mathcal{A}_N^s)^{\frac{1}{4}-\epsilon}(I - P_N)[y_0 - Dg_N(0)]$

$$- \int_0^t (-\mathcal{A}_N^s)^{-(\frac{1}{4}-\epsilon)} e^{\mathcal{A}_N^s(t-\tau)}(-\mathcal{A}_N^s)^{\frac{1}{4}-\epsilon}(I - P_N)D\dot{g}_N(\tau)d\tau$$

$$\in C\left([0,T]; \mathcal{D}((-\mathcal{A}_N^s)^{\frac{1}{4}-\epsilon})\right) \cap L^2\left(0,\infty; \mathcal{D}((-\mathcal{A}_N^s)^{\frac{3}{4}-\epsilon})\right),$$

as in the proof of Lemma 3.1.2(i), under assumption (3.1.15). Since $\dot{g}_N = \text{Re } \dot{v}_N$ is exponentially decaying, see (B.2.3), we can take $T = \infty$. Hence, recalling $y = \eta + Dg_N$ from (B.2.4), we have:

(B.2.12) $(I - P_N)y(t) = (I - P_N)\eta(t) + (I - P_N)Dg_N(t)$

$$\in C\left([0,T]; \mathcal{D}((-\mathcal{A}_N^s)^{\frac{1}{4}-\epsilon} \equiv (H^{\frac{1}{2}-2\epsilon}(\Omega))^d \cap H\right)$$

$$\cap L^2\left(0,\infty; \mathcal{D}((-\mathcal{A}_N^s)^{\frac{3}{4}-\epsilon} \equiv (H^{\frac{3}{2}-2\epsilon}(\Omega))^d \cap V\right),$$

see (1.16), since from (B.2.1) and (3.1.3d) with $s = \frac{1}{2}$, we have $Dg_N(t) \in C^n(0, \infty; (H^1(\Omega))^d \cap H)$. Then (B.2.10) yields (B.2.6). □

B.3. Completion of the proof of Theorem 2.5 and Theorem 2.6 for the N-S model (1.1), $d = 2$

We can complete the proof of Theorem 2.5 and Theorem 2.6 for the N-S model (1.1), $d = 2$, as follows.

Completion of proof of Theorem 2.5. In contrast with Section 4 for $d = 3$, here we consider the Optimal Control Problem

(B.3.1) $$J(u, y) = \int_0^\infty |y(t)|_H^2 + |u(t)|_U^2 dt,$$

with corresponding state and output spaces

$$W_1 \equiv (H^{\frac{1}{2}-\epsilon}(\Omega))^2 \cap H \text{ (free dynamics)}; \quad Z_1 \equiv (H^{\frac{3}{2}-\epsilon}(\Omega))^2 \cap H \text{ (output space)}$$

[definitely more amenable than the spaces W, Z in (4.1.1b) which have to be considered for $d = 3$]. Then, Theorem B.1.1 yields that the Finite Cost Condition is satisfied for this (more amenable) OCP. Then, we can invoke the corresponding Optimal Control Theory and corresponding Algebraic Riccati Equations, as already available in the literature for about 20 years [**L-T.1**, Sections 5.1 and 6.1, pp. 36–52], [**L-T.2**, Vol. 1, Thm. 3.2.1(ii), p. 187], to be refined as in the subsequent appendix B.4, since the operator $\mathcal{A}$ in (1.11) is a benign $\frac{1}{2}$-relatively bounded perturbation of the self-adjoint operator $\nu_0 A_0$. Then (see [**L-T.2**, Vol. 2, Thm. 2.2.1(a_3), p. 126]; and Appendix 2B, p. 168–170], for the self-adjoint case), it is the ARE

(B.3.2a) $(\mathcal{A}^* \tilde{R} x, y)_H + (\tilde{R} \mathcal{A} x, y)_H + (x, y)_H = \langle D^* \mathcal{A}^* \tilde{R} x, D^* \mathcal{A}^* \tilde{R} y \rangle_U$

(B.3.2b) $$= \left\langle \frac{\partial \tilde{R} x}{\partial \nu}, \frac{\partial \tilde{R} y}{\partial \nu} \right\rangle_U, \forall x, y \in H$$

[far more desirable than the ARE (4.5.1) for $d = 3$] that generates the Riccati operator $\tilde{R}$ as its unique positive, self-adjoint solution. This follows from the above-quoted references [**L-T.1**], [**L-T.2**] (which report on work of their authors of the late 70's–mid 80's), together with the refinement of the forthcoming Appendix B.4 (inspired by the above references). Thus, Section 4 can be entirely dispensed with. The final result of Theorem 2.5 then follows in the same way as Theorem 2.3 follows from Theorem 2.1 (in this case using Section 4), via the counterpart of Section 6.

Completion of proof of Theorem 2.6. Same as the proof of Theorem 2.5 above, this time using Theorem B.2.1, instead of B.1.1, to satisfy the Finite Cost condition of the aforementioned OCP within the class of finite-dimensional controllers having arbitrary support Γ_1 on $\Gamma = \partial\Omega$, under the FDSA = (3.6.2).

B.4. A regularity property of the Riccati operator corresponding to the linearized operator $\mathcal{A}$ in (1.11)

The following abstract setting includes, in particular, the case of the linearized operator $\mathcal{A}$ defined in (1.11). The present Appendix B.4 may be considered as providing an additional physically important illustration, over those already included in [**L-T.2**, Chapter 2, Appendix 2B, p. 168–170], where the regularity of the Riccati operator noted in [**L-T.2**, Chapter 2, Theorem 2.2.1(a_3)]—and recalled in (B.4.9) below—can be further "improved by ϵ" as to achieve (B.4.10).

Abstract setting. In this Appendix B.4, we embrace the abstract parabolic setting of [**L-T.2**, Chapter 2]. Thus, we consider the abstract state equation

(B.4.1) $$\dot{y} = \mathcal{A}y + \mathcal{B}u \in [\mathcal{D}(\mathcal{A}^*)]', \quad y(0) = y_0 \in Y,$$

where (i) the Hilbert spaces Y and U are, respectively, the state space and the control space; (ii) $\mathcal{A}$ is the infinitesimal generator of a s.c. analytic semigroup $e^{\mathcal{A}t}$ on Y, $t \geq 0$; (iii) $(\lambda - \mathcal{A})^{-\gamma}\mathcal{B} \in \mathcal{L}(U;Y)$, for some constant $0 < \gamma < 1$, $\lambda \in \rho(\mathcal{A})$ = the resolvent set of $\mathcal{A}$. In (B.4.1), [ ]' denotes the dual space of $\mathcal{D}(\mathcal{A}^*) \subset Y$ with respect to Y as a pivot space. We note explicitly that we are considering the general case (as illustrated by the linearized N-S operator (1.11)), where the s.c. analytic semigroup $e^{\mathcal{A}t}$ is unstable on Y. Thus, as in [**L-T.2**, Chapter 2], we then introduce the translated generator

(B.4.2a) $$-\hat{\mathcal{A}} \equiv \mathcal{A} - \omega I, \quad \omega > 0,$$

with constant $\omega > 0$ sufficiently large so that the corresponding s.c. analytic semigroup $e^{-\hat{\mathcal{A}}t}$ is exponentially stable

(B.4.2b) $$\|e^{-\hat{\mathcal{A}}t}\|_{\mathcal{L}(Y)} \leq Me^{-at}, \quad t \geq 0, \ a > 0,$$

and hence the fractional powers $\hat{\mathcal{A}}^\sigma$, $0 \leq \sigma \leq 1$, of $\hat{\mathcal{A}}$ are well defined.

We next impose the following two additional hypotheses:
(1) We assume that the generator $\mathcal{A}$ decomposes as

(B.4.3a) $$\begin{cases} \mathcal{A} \equiv -A + A_0, \ \mathcal{A}^* = -A + A_0^*, \\ Y \supset \mathcal{D}(\mathcal{A}) = \mathcal{D}(\mathcal{A}^*) = \mathcal{D}(A) = \mathcal{D}(\hat{\mathcal{A}}) \to Y; \end{cases}$$
(B.4.3b)

$$\text{(B.4.4a)} \quad \begin{cases} A = \text{strictly positive, self-adjoint operator on } Y; \\ \text{(B.4.4b)} \quad \mathcal{D}(A_0) = \mathcal{D}(A_0^*) = \mathcal{D}(A^\alpha) = \mathcal{D}(\hat{\mathcal{A}}^\alpha), \text{ for some fixed } 0 \leq \alpha < 1, \\ \text{thus} \\ \text{(B.4.4c)} \quad A_0 A^{-\alpha}, A_0 \hat{\mathcal{A}}^{-\alpha}, A_0^* A^{-\alpha}, A_0^* \hat{\mathcal{A}}^{-\alpha} \in \mathcal{L}(Y); \\ \text{hence } A^{-\alpha} A_0^*, \hat{\mathcal{A}}^{*-\alpha} A_0^*, A^{-\alpha} A_0, \hat{\mathcal{A}}^{*-\alpha} A_0 \in \mathcal{L}(Y) \text{ by adjointness.} \end{cases}$$

(2) The observation operator (denoted by R in [**L-T.2**, Chapter 2]) is the identity operator on Y (see (B.4.6) below).

Thus, by (1), the generator $\mathcal{A}$ is the sum of a strictly negative, self-adjoint operator and a lower-order term.

Remark B.4.1. The case of the linearized operator $\mathcal{A}$ in (1.11) is included in the above assumption (1) when $\alpha = \frac{1}{2}$, see (1.10c), where, moreover, Y is the space H in (1.5a) and $\gamma = \frac{3}{4} + \epsilon$, or $1 - \gamma = \frac{1}{4} - \epsilon$, $\forall\, \epsilon > 0$, by virtue of relation (3.1.3e).

We next consider the following Optimal Control Problem (OCP), associated with the dynamics (B.4.1), over an infinite time horizon:

$$\text{(B.4.5)} \qquad \min J(u, y) \text{ over all } u \in L^2(0, \infty; U);$$

$$\text{(B.4.6)} \qquad J(u, y) \equiv \int_0^\infty \left[\|y(t; y_0)\|_Y^2 + \|u(t)\|_U^2 \right] dt, \quad y_0 \in Y.$$

Under the additional Finite Cost condition (that is, the functional (B.4.6) is "proper" for all $y_0 \in Y$, in the language of optimization theory, then the main abstract parabolic result in [**L-T.2**, Theorem 2.2.1, p. 125] *applies to the OCP (B.4.5), (B.4.6) for (B.4.1)*. In particular, there exists a unique optimal pair $\{u^0(\,\cdot\,; y_0), y^0(\,\cdot\,; y_0)\}$. Moreover, if we denote by P (in this Appendix B.4) the corresponding Riccati operator

$$\text{(B.4.7)} \qquad (Px, x)_Y = \inf_{u \in L^2(0, \infty; U)} J(u, y) = J(u^0, y^0) = J^0(y_0), \quad y_0 \in Y,$$

so that $P = P^* > 0$ is a positive, self-adjoint operator on Y, we then have

$$\text{(B.4.8)} \qquad \hat{\mathcal{A}}^{*\theta} P, \ \hat{\mathcal{A}}^\theta P, \ A^\theta P \in \mathcal{L}(Y), \text{ for all } 0 \leq \theta < 1$$

[**L-T.2**, Theorem 2.1.2(a_3), p. 126]. In writing (B.4.8), we have also used

$$\text{(B.4.9)} \qquad \mathcal{D}(\hat{\mathcal{A}}^\theta) = \mathcal{D}(\hat{\mathcal{A}}^{*\theta}) = \mathcal{D}(A^\theta), \quad 0 \leq \theta \leq 1,$$

as it follows from (B.4.3a) by interpolation.

Remark B.4.2. In the case where $A_0 = 0$ in (B.4.3), or more generally, in the case where $\mathcal{A}$ is a stable normal operator (so that $\mathcal{D}(\mathcal{A}) = \mathcal{D}(\mathcal{A}^*)$ is automatically true), [**L-T.2**, Appendix 2B, p. 168–170] shows then that we can take $\theta = 1$ in (B.4.8). These results *cannot be directly invoked in the present setting*, as $\mathcal{A}$ is neither normal nor stable. Nevertheless, we shall suitably modify and complement the arguments of [**L-T.2**, Appendix 2B, p. 168–170] to obtain the desired extension $\theta = 1$ in (B.4.8), also in our present setting (which includes the linearized Navier-Stokes operator $\mathcal{A}$ in (1.11), see Remark B.4.1). This conclusion $\theta = 1$ is shown in the following result.

B.4. A REGULARITY PROPERTY OF THE RICCATI OPERATOR

PROPOSITION B.4.1. *Consider the dynamics (B.4.1) subject to the standard parabolic assumption (i), (ii), (iii) as stated below (B.4.1), adn the additional more specific assumptions (1) = (B.4.3), (2) for the corresponding OCP. Let the Finite Cost Condition hold true. Then, [**L-T.2**, Theorem 2.2.1, p. 125] holds true where:*

(a) its part (a_3) is improved as follows:

(B.4.10) $$\mathcal{A}P, \; \hat{\mathcal{A}}^*P, \hat{\mathcal{A}}P \in \mathcal{L}(Y)$$

(thus extending the validity of (B.4.8) to include the case $\theta = 1$). The corresponding ARE, where $\mathcal{B}P \in \mathcal{L}(Y)$, is given by

(B.4.11) $$\mathcal{A}^*P + P\mathcal{A} + I = (\mathcal{B}^*P)^*\mathcal{B}^*P$$

*(an ϵ-improvement over the general case of [**L-T.2**, Theorem 2.2.1, Eqn. (2.2.2), p. 125].*

*(b) Let $A_F = (\mathcal{A} - \mathcal{B}\mathcal{B}^*P)$ be, as in [**L-T.2**, Theorem 2.2.1] the generator of the corresponding s.c. analytic feedback semigroup $e^{A_F t}$ on Y, so that the optimal solution is $y^0(t;y_0) \equiv e^{A_F t}y_0$, $y_0 \in Y$. Then, the operator PA_F can be extended by continuity to a bounded operator in $\mathcal{L}(Y)$:*

(B.4.12) $$PA_F \in \mathcal{L}(Y), \; \text{hence} \; A_F^*P \in \mathcal{L}(Y).$$

(c) With $0 < \gamma < 1$ the constant postulated in (iii) below (B.4.1) (e.g., $\gamma = \frac{3}{4} + \epsilon$ for the linearized N-S operator in (1.11), see Remark B.4.1), we then have that

(B.4.13a) $$\left\{ \begin{array}{l} P: \; \text{continuous} \; \mathcal{D}(\hat{\mathcal{A}}^{1-\gamma}) \to \mathcal{D}(\hat{\mathcal{A}}^{2-\gamma}) \\ \text{equivalently,} \; \hat{\mathcal{A}}^{2-\gamma} P \hat{\mathcal{A}}^{-(1-\gamma)} \in \mathcal{L}(Y). \end{array} \right.$$

(B.4.13b)

Proof. Part (a). We suitably modify and complement the arguments of [**L-T.2**, Appendix 2B, pp. 168–170].

We recall from [**L-T.2**, Theorem 2.2.1(a_8), Eqn. (2.2.9), p. 127] that the Riccati operator P satisfies the relationship

(B.4.14) $$Px = \int_0^\infty e^{-\hat{\mathcal{A}}^* t}[I + 2\omega P]e^{-\omega t}e^{A_F t}x \, dt, \quad x \in Y,$$

where, recalling

(B.4.15) $$\dot{y}^0 = A_F y^0 = (\mathcal{A} - \mathcal{B}\mathcal{B}^*P)y^0 = \mathcal{A}y^0 + \mathcal{B}u^0,$$

and (B.4.2a), we have [**L-T.2**, p. 125–126; Eqn. (2.3.3.18), p. 137],

(B.4.16a) $$\hat{y}^0(t;x) = e^{-\omega t}y^0(t;x) = e^{-\omega t}e^{A_F t}x = e^{-\hat{\mathcal{A}}t}x + \{\hat{L}\hat{u}^0(\,\cdot\,;x)\}(t);$$

(B.4.16b) $$\{\hat{L}\hat{u}^0(\,\cdot\,;x)\}(t) = \int_0^t e^{-\hat{\mathcal{A}}(t-\tau)}\mathcal{B}\hat{u}^0(\tau;x)\,d\tau; \quad \hat{u}^0(t;x) = e^{-\omega t}u^0(t;x);$$

Substituting (B.4.16a) into (B.4.14) yields for $x \in Y$:

(B.4.17a) $$Px = I_1 x + I_2 x + I_3 x;$$

(B.4.17b) $$I_1 x = \int_0^\infty e^{-\hat{\mathcal{A}}^* t}e^{-\hat{\mathcal{A}}t}x \, dt; \quad I_2 x = 2\omega \int_0^\infty e^{-\hat{\mathcal{A}}^* t}Pe^{-\hat{\mathcal{A}}t}x \, dt;$$

(B.4.18) $$I_3 x = \int_0^\infty e^{-\hat{\mathcal{A}}^* t}[I + 2\omega P]\{\hat{L}\hat{u}^0(\,\cdot\,;x)\}(t)dt.$$

Term I_3. We shall show that

(B.4.19) $$\hat{\mathcal{A}}^* I_3 \in \mathcal{L}(Y); \text{ or } AI_3, \ \mathcal{A}I_3 \in \mathcal{L}(Y).$$

To this end, we use the same argument as the one in [**L-T.2**, Lemma 2B.1, p. 169] for the first subterm of I_3, while, for its second subterm, we use in addition the *a-priori* knowledge (B.4.8). All this is based on the following property contained in Theorem 3.1.8, Eqn. (3.1.70):
(B.4.20)
$$\hat{\mathcal{A}}^{(\frac{1}{4}-\epsilon)}\{\hat{L}\hat{u}^0(\,\cdot\,;x)\}(t), \ \hat{\mathcal{A}}^{*(\frac{1}{4}-\epsilon)}\{\hat{L}\hat{u}^0(\,\cdot\,;x)\}(t) \in L^2(0,\infty;Y) \cap C([0,T_0];Y), \ x \in Y,$$

continuously on $x \in Y$, for all $0 < T_0 < \infty$, $\epsilon > 0$. Thus, we rewrite (B.4.18) more conveniently, for any $\epsilon > 0$ and any $\sigma > 0$, both arbitrarily small, as

$$\text{(B.4.21)} \ \hat{\mathcal{A}}^* I_3 x = \int_0^\infty (\hat{\mathcal{A}}^{*\frac{3}{4}+\epsilon} e^{-\hat{\mathcal{A}}^* t})[\hat{\mathcal{A}}^{*(\frac{1}{4}-\epsilon)}\{\hat{L}\hat{u}^0(\,\cdot\,;x)\}(t)] dt$$

$$+ 2\omega \int_0^\infty (\hat{\mathcal{A}}^{*\sigma} e^{-\hat{\mathcal{A}}^* t})(\hat{\mathcal{A}}^{*1-\sigma} P)\{\hat{L}\hat{u}^0(\,\cdot\,;x)\}(t) dt, \ x \in Y.$$

We split $\int_0^\infty = \int_0^1 + \int_1^\infty$. We then use the standard analytic semigroup estimate $\mathcal{O}(e^{-at}/t^{\frac{3}{4}+\epsilon})$, $a > 0$, see (B.4.2b), for the term ( ) in the first integral $\int_0^1$ of (B.4.21) along with the continuity property $C([0,1];Y)$ in (B.4.20) for the second term [ ] in the integral $\int_0^1$; and the $L_2^2(0,\infty;Y)$-property in (B.4.20) for the second term [ ] in the integral $\int_1^\infty$. In the second integral $\int_1^\infty$ of (B.4.21), we also use (B.4.8) for P, with $\theta = 1 - \sigma < 1$. This way, the analysis on (B.4.21) leads to (B.4.19) as $\hat{\mathcal{A}}^* I_3 \in \mathcal{L}(Y)$.

Term I_1. We shall show that

(B.4.22) $$AI_1 \in \mathcal{L}(Y); \text{ or } \hat{\mathcal{A}}^* I_1, \ \hat{\mathcal{A}} I_1 \in \mathcal{L}(Y).$$

We shall use the critical property that, in the present case (B.4.3), (B.4.4), with $0 \le \alpha < 1$, we have

(B.4.23) $$e^{-\hat{\mathcal{A}}t}, e^{-\hat{\mathcal{A}}^* t} : \text{continuous } Y \to L^2(0,\infty; \mathcal{D}(\hat{\mathcal{A}}^{\frac{1}{2}}) = \mathcal{D}(\hat{\mathcal{A}}^{*\frac{1}{2}}) = \mathcal{D}(A^{\frac{1}{2}})).$$

The validity of the regularity properties (B.4.23) in the case of the linearized N-S operator (1.11), where then $\alpha = \frac{1}{2}$, was already noted e.g., [**B-T.1**, Appendix A] and tacitly used e.g., in (3.1.8), in the proof of Proposition 3.1.9; etc. The regularity (B.4.23) continues to hold true for the operator in (B.4.3), (B.4.4) as long as $0 \le \alpha < 1$. (This is seen e.g., by energy methods: multiply the equation $\dot{y} = -\hat{\mathcal{A}}y$ by y and integrate by parts.) To prove (B.4.22), say initially, $\hat{\mathcal{A}}^* I_1 \in \mathcal{L}(Y)$, we take

B.4. A REGULARITY PROPERTY OF THE RICCATI OPERATOR

$x, y \in Y$ and estimate

$$\text{(B.4.24)} \quad |(\hat{\mathcal{A}}^* I_1 x, y)_Y| = \left|\left(\int_0^\infty \hat{\mathcal{A}}^* e^{-\hat{\mathcal{A}}^* t} e^{-\hat{\mathcal{A}} t} x \, dt, y\right)_Y\right|$$

$$\text{(B.4.25)} \qquad = \left|\left(\int_0^\infty \hat{\mathcal{A}}^{*\frac{1}{2}} e^{-\hat{\mathcal{A}}^* t} (\hat{\mathcal{A}}^{*\frac{1}{2}} \hat{\mathcal{A}}^{-\frac{1}{2}}) \hat{\mathcal{A}}^{\frac{1}{2}} e^{-\hat{\mathcal{A}} t} x \, dt, y\right)_Y\right|$$

$$\text{(B.4.26)} \qquad = \left|\left(\int_0^\infty (\hat{\mathcal{A}}^{*\frac{1}{2}} \hat{\mathcal{A}}^{-\frac{1}{2}}) \hat{\mathcal{A}}^{\frac{1}{2}} e^{-\hat{\mathcal{A}} t} x, \hat{\mathcal{A}}^{\frac{1}{2}} e^{-\hat{\mathcal{A}} t} y\right)_Y dt\right|$$

$$\text{(B.4.27)} \quad \text{(by (B.4.23))} \;\; \leq \;\; C|x|_Y |y|_Y, \quad \forall x, y \in Y,$$

by using, in the last step: Schwarz inequality, the property $(\hat{\mathcal{A}}^{*\frac{1}{2}} \hat{\mathcal{A}}^{-\frac{1}{2}}) \in \mathcal{L}(Y)$ (from $\mathcal{D}(\hat{\mathcal{A}}) = \mathcal{D}(\hat{\mathcal{A}}^*)$ and interpolation, with $\alpha < 1$), and finally (B.4.23). By (B.4.27), we conclude that $\hat{\mathcal{A}}^* I_1 \in \mathcal{L}(Y)$, as desired. The other properties in (B.4.22) follow via (B.4.3b).

Term I_2. Because of property (B.4.8), on P, this is better behaved than I_1. We obtain $\hat{\mathcal{A}}^* I_2 \in \mathcal{L}(Y)$.

Conclusion of proof of Part (a). Combining (B.4.22) on $\hat{\mathcal{A}}^* I_1$, (B.4.19) on $\hat{\mathcal{A}}^* I_3, \ldots$, and $\hat{\mathcal{A}}^* I_2$ above in (B.4.17a) yields $\hat{\mathcal{A}}^* P \in \mathcal{L}(Y), \ldots$, and (B.4.10) is proved.

Result (B.4.10) allows us to refine, in the present case, the general result [**L-T.2**, Theorem 2.2.1, p. 125], where $\mathcal{B}^* P = \mathcal{B}^* \mathcal{A}^{*-\gamma} \mathcal{A}^{*\gamma} P \in \mathcal{L}(Y; U)$, γ = constant below (B.4.1), and obtain the ARE (B.4.11).

(b) We recall from [**F-L-T.1**, formula (4.61)]

$$\text{(B.4.28)} \qquad \mathcal{A}^* P x = -x - P A_F x \in Y, \text{ originally for } x \in \mathcal{D}(A_F).$$

Thus, $-PA_F$ has a bounded extension in $\mathcal{L}(Y)$ given by $[I + \mathcal{A}^* P]$, by use of (B.4.28).

(c) The proof of this part is similar to that of part (a). Proposition B.4.1 is proved. □

APPENDIX C

Equivalence between unstable and stable versions of the Optimal Control Problem of Section 4

The present Appendix C provides a useful *equivalence* in the study of optimal control problems with quadratic cost functionals: between the original *unstable* dynamics, assumed to satisfy, of course, the Finite Cost Condition (FCC) and a suitable *stable* version, which then *a-fortiori* satisfies the FCC. Thus, without loss of generality, the former unstable problem may be reduced to the latter stable one, where then a more streamlined and, above all, explicit treatment is possible. After establishing this equivalence here, we shall then take advantage of this useful reduction—from the unstable to the stable case—in the present study of the OCP of the *linearized* Navier-Stokes equation beginning with Section 4.3 and throughout Section 4.5. That the present OCP (for the more challenging case $d = 3$) does satisfy the critical FCC was established in Section 3.5 in full generality; and in Sections 3.6–3.7 under the FDSA = (3.6.2), with a more desirable (open-loop) boundary control.

Setting of Section 4.1. We have introduced the state space W of the (closed loop) optimal dynamics, the control space U, and the observation space Z, where we recall from (4.1.1) that

$$W \equiv \left(H^{\frac{1}{2}+\epsilon}(\Omega)\right)^d \cap H; \quad U = \{u \in (L^2(\Gamma))^d : u \cdot \nu = 0\}; \quad Z \equiv \left(H^{\frac{3}{2}+\epsilon}(\Omega)\right)^d \cap H,$$

with space H defined in (1.5a). We have let so far:

(i) $\mathcal{A}$ be the infinitesimal generator of a s.c. semigroup $e^{\mathcal{A}t}$, $t \geq 0$, on H, assumed to be unstable. (Analyticity of $e^{\mathcal{A}t}$ is not strictly needed.)

(ii) $\mathcal{B}$ ($= -\mathcal{A}D$) be a linear operator $U \to [\mathcal{D}(\mathcal{A}^*)]'$ (adjoint and duality with respect to the pivot space H); equivalently, $R(\lambda_0, \mathcal{A})\mathcal{B} \in \mathcal{L}(U; W)$, for points λ_0 in the resolvent of $\mathcal{A}$.

OCP #1. The original problem consists of the dynamics:

(C.1) $$y_t = \mathcal{A}y + \mathcal{B}u; \quad y(0) = y_0 \in W;$$

with cost function (4.1.3),

(C.2a) $$J(u, y) \equiv \int_0^\infty [|Qy(t)|_H^2 + |u(t)|_U^2] dt$$

to be minimized:

(C.2b) $$J^0(y_0) \equiv \inf_{u \in L^2(0,\infty;U)} \int_0^\infty [|Qy(t)|_H^2 + |u(t)|_U^2] dt,$$

where Q: isomorphism $(H^{\frac{3}{2}+\epsilon}(\Omega))^d \cap H$ onto H, under the *assumption* of the FCC. This assumptions was verified to hold true for $d = 3$ in Section 3.5 in the general case; and in Sections 3.6–3.7 under the FDSA = (3.6.2). Accordingly, by Propositions 4.1.1, 4.1.3, and Corollary 4.1.2, there exists a unique optimal pair $\{u^0(t; y_0), y^0(t; y_0)\}$ and a non-negative, unbounded, self-adjoint operator R on H, satisfying, moreover, $R \in \mathcal{L}(W; W')$, such that

$$
\begin{aligned}
\text{(C.3)} \quad J^0(y_0) &= J(u^0(\,\cdot\,; y_0), y^0(\,\cdot\,; y_0)) \\
&= (Ry_0, y_0)_H = \int_0^\infty [|Qy^0(t; y_0)|_H^2 + |u^0(t; y_0)|_U^2] dt,
\end{aligned}
$$

recall (4.1.4), from which it readily follows via the semigroup property (4.2.5), (4.2.6) of the optimal pair that [**T.3**]

$$
\text{(C.4)} \quad (Ry^0(t; y_0), y^0(t; y_0))_H = \int_t^\infty [|Qy^0(\tau; y_0)|_H^2 + |u^0(\tau; y_0)|_U^2] d\tau, \quad t \geq 0.
$$

OCP #2. We consider the translated dynamics

$$
\text{(C.5a)} \quad z_t = \mathcal{A}_\lambda z + \mathcal{B}\mu; \quad z(0) = y_0 \in W;
$$

$$
\text{(C.5b)} \quad \mathcal{A}_\lambda \equiv (\mathcal{A} - \lambda I), \text{ for a sufficiently large } \lambda > 0,
$$

so that $e^{\mathcal{A}_\lambda t}$ is exponentially uniformly stable on H, along with the cost functional

$$
\text{(C.5c)} \quad J_\lambda(\mu, z) \equiv \int_0^\infty [|Qz(t)|_H^2 + 2\lambda(Rz(t), z(t))_H + |\mu(t)|_U^2] dt,
$$

to be likewise minimized

$$
\text{(C.6)} \quad J_\lambda^0(y_0) \equiv \inf_{\mu \in L^2(0,\infty;U)} \int_0^\infty [|Qz(t)|_H^2 + 2\lambda(Rz(t), z(t))_H + |\mu(t)|_U^2] dt,
$$

where R is defined in (C.3). Since $\lambda > 0$ and the FCC is true for (C.1), (C.2), then the corresponding FCC for (C.5a–c) is *a-fortiori* satisfied. In fact, let $\{\bar{u}(t), \bar{y}(t; y_0)\}$ $y_0 \in W$, be a compatible pair for the dynamics (C.1), such that $J(\bar{u}, \bar{y}) < \infty$, as the FCC holds for (C.1) [as already established in Section 3.5]. Then the pair $\{\bar{u}_\lambda(t) = e^{-\lambda t}\bar{u}(t), \bar{y}_\lambda(t; y_0) = e^{-\lambda t}\bar{y}(t; y_0)\}$ is a compatible pair for the dynamics (C.5a), see (C.15a) below. Moreover, we have $J_\lambda(\bar{u}_\lambda, \bar{y}_\lambda) < \infty$, with $\bar{u}_\lambda \in L^2(0, \infty; U)$ *a-fortiori*, since, by (4.1.8), we have that $R^{\frac{1}{2}}$ is dominated by Q: $(Rz(t), z(t))_H \leq C|z(t)|_W^2 \leq C|z(t)|_Z^2 = C|Qz(t)|_H^2$, and thus

$$
J_\lambda(\mu, z) \leq \int_0^\infty \{[1 + 2\lambda C]|Qz(t)|_H^2 + |\mu(t)|^2\} dt \leq (1 + 2\lambda C) J(\mu, z).
$$

Hence the FCC is a *a-fortiori* verified for (C.5a)–(C.6) as a consequence of the FCC being satisfied for (C.1)-(C.2). Thus there exists a unique optimal pair $\{\mu_\lambda^0(t; y_0), z_\lambda^0(t; y_0)\}$ and a non-negative, unbounded, self-adjoint operator R_λ on W, with $R_\lambda \in \mathcal{L}(W; W')$ such that

$$
\begin{aligned}
\text{(C.7)} \quad J_\lambda^0(y_0) &= J_\lambda(\mu_\lambda^0(\,\cdot\,; y_0), z_\lambda^0(\,\cdot\,; y_0)) = (R_\lambda y_0, y_0)_H \\
&= \int_0^\infty [|Qz_\lambda^0(t; y_0)|_H^2 + 2\lambda(Rz_\lambda^0(t; y_0), z_\lambda^0(t; y_0))_H + |\mu_\lambda^0(t; y_0)|^2] dt \\
\text{(C.8)} \quad &= \int_0^\infty \{([Q^*Q + 2\lambda R]z_\lambda^0(t; y_0))_H + |\mu_\lambda^0(t; y_0)|_U^2\} dt
\end{aligned}
$$

from which it readily follows [**T.3**] that

(C.9) $(R_\lambda z_\lambda^0(t; y_0), z_\lambda^0(t; y_0))_H$

$$= \int_t^\infty \{([Q^*Q + 2\lambda R]z_\lambda^0(\tau; y_0), z_\lambda^0(\tau; y_0))_H + |\mu_\lambda^0(\tau; y_0)|_U^2\} dt.$$

It is well-known [**F-L-T.1**], [**L-T.1**], [**L-T.2**], that R_λ is given explicitly by

$$R_\lambda y_0 = \int_0^\infty e^{\mathcal{A}_\lambda^* t}[Q^*Q + 2\lambda R]z_\lambda^0(t; y_0) dt.$$

The main goal of this Appendix C is the following *equivalence* theorem between the OCP#1 and the OCP#2.

THEOREM C.1. *Consider the unstable OCP #1 under the FCC, and the corresponding stable OCP #2. Then, the following equivalence holds true:*

(i) the optimal pair $\{u^0(t; y_0), y^0(t; y_0)\}$ of OCP #1 and the optimal pair $\{\mu_\lambda^0(t; y_0), z_\lambda^0(t; y_0)\}$ of OCP#2 are related by the following identities:

(C.10) $$u^0(t; y_0) = \mu_\lambda^0(t; y_0)e^{\lambda t};$$

$$S(t)y_0 \equiv y^0(t; y_0) = z_\lambda^0(t; y_0)e^{\lambda t} \equiv S_\lambda(t)e^{\lambda t}y_0, \ t \geq 0;$$

(ii) the two Riccati operators R and R_λ of the two OCP's coincide:

(C.11) $$R = R_\lambda,$$

so that the optimal costs of the two problems coincide:

(C.12) $$J^0(y_0) = J_\lambda^0(y_0), \ y_0 \in W; \ that \ is, \ explicitly,$$

$$J(u^0(\cdot; y_0), y^0(\cdot; y_0)) = J_\lambda(\mu_\lambda^0(\cdot; y_0), z_\lambda^0(\cdot; y_0));$$

(iii) (counterpart of (4.3.21))

(C.13a) $$u^0(t; 0; y_0)e^{-\lambda t} = -V_{t,\lambda}^* y_0,$$

where

(C.13b) $$V_{t,\lambda}^* x = \mathcal{B}^* \int_t^\infty e^{\mathcal{A}_\lambda^*(\tau-t)}[Q^*Q + 2\lambda R]S_\lambda(\tau)x \, d\tau;$$

(iv) recalling (3.1.28) [with $\mathcal{B}^* = -D^*\mathcal{A}^*$] (counterpart of (4.3.13))

(C.14) $$V_{t,\lambda}^* = L_{t,\lambda}^*[Q^*Q + 2\lambda R]S_\lambda(\cdot): \ continuous \ \mathcal{D}(A_R^2) \to H^{-1-\epsilon}(\Gamma).$$

We shall provide below two proofs of the key Steps 4-5 needed to establish this important result. The first proof has the advantage of being purely 'algebraic.' The second one has the advantage that it would apply also to a more general non-linear cost.

First Proof. (ii) The key of this result is the coincidence $R = R_\lambda$ of the Riccati operators.

Step 1. It is elementary to verify that:

(C.15a) $\begin{cases} \{u, y\} \text{ is a compatible pair for (C.1); that is, } y_t = \mathcal{A}y + \mathcal{B}u, \\ \text{and hence } \{u, y\} \text{ is a competitive pair for OCP \#1 if and only if} \\ \{u_\lambda \equiv e^{-\lambda t}u, \ y_\lambda \equiv e^{-\lambda t}y\} \text{ is a compatible pair for (C.5a); that} \\ \text{is } \dot{y}_\lambda = \mathcal{A}_\lambda y_\lambda + \mathcal{B}u_\lambda, \text{ and hence } \{u_\lambda, y_\lambda\} \text{ is a competitive pair for} \\ \text{OCP \#2.} \end{cases}$

Indeed, we multiply (C.1) by $e^{-\lambda t}$, use $e^{-\lambda t}y_t = (e^{-\lambda t}y)_t + \lambda(e^{-\lambda t}y)$, to establish the "if part," and, conversely, for the "only if" part.

Step 2. With $\{u^0(t; y_0), y^0(t; y_0)\}$ the optimal pair of OCP #1, introduce the pair

(C.15b) $\qquad \tilde{u}_\lambda(t; y_0) \equiv e^{-\lambda t}u^0(t; y_0); \quad \tilde{y}_\lambda(t; y_0) \equiv e^{-\lambda t}y^0(t; y_0), \ t \geq 0.$

By Step 1, $\{\tilde{u}_\lambda, \tilde{y}_\lambda\}$ is a compatible pair of (C.5a), hence a competitive pair for OCP #2. [There is no claim, at this stage, that $\{\tilde{u}_\lambda, \tilde{y}_\lambda\}$ is the optimal pair of OCP #2.] A key step in the overall proof consists in establishing that

$$\text{(C.16)} \quad J(u^0, y^0) = (Ry_0, y_0)_H = \int_0^\infty \left[|Qy^0(t; y_0)|_H^2 + |u^0(t; y_0)|_U^2\right] dt$$

$$= \int_0^\infty \left[|Q\tilde{y}_\lambda(t; y_0)|_H^2 + |\tilde{u}_\lambda(t; y_0)|_U^2 + 2\lambda(R\tilde{y}_\lambda(t; y_0), \tilde{y}_\lambda(t; y_0))_H\right] dt$$

$$= J_\lambda(\tilde{u}_\lambda, \tilde{y}_\lambda).$$

PROOF (C.16). We begin with one part of the RHS of (C.16): by (C.4), setting

(C.17) $\qquad g(t) \equiv (Ry^0(t; y_0), y^0(t; y_0))_H$

$$\equiv \int_t^\infty \left[|Qy^0(\tau; y_0)|_H^2 + |u^0(\tau; y_0)|_U^2\right] d\tau, \quad g(\infty) = 0,$$

we compute via (C.15b),

$$\text{(C.18)} \quad \int_0^\infty 2\lambda(R\tilde{y}_\lambda(t; y_0), \tilde{y}_\lambda(t; y_0))_H dt$$

$$= 2\lambda \int_0^\infty e^{-2\lambda t}(Ry^0(t; y_0), y^0(t; y_0))_H dt$$

(C.19) (by (C.17)) $\displaystyle = \int_0^\infty \frac{d}{dt}\left(e^{-2\lambda t}\right)(-g(t))dt = \left[-e^{-2\lambda t}g(t)\right]_{t=0}^{t=\infty}$

$$+ \int_0^\infty e^{-2\lambda t}\frac{d}{dt}g(t)dt$$

(C.20) (by (C.17)) $\displaystyle = g(0) - \int_0^\infty e^{-2\lambda t}\left[|Qy^0(t; y_0)|_H^2 + |u^0(t; y_0)|_U^2\right] dt$

(C.21) $\displaystyle = \int_0^\infty \left[|Qy^0(t; y_0)|_H^2 + |u^0(t; y_0)|_U^2\right] dt$

$$- \int_0^\infty \left[|Q\tilde{y}_\lambda(t; y_0)|_H^2 + |\tilde{u}_\lambda(t; y_0)|_U^2\right] dt,$$

invoking, in the last step, (C.17) for $g(0)$ and (C.15b) for $\{\tilde{u}_\lambda, \tilde{y}_\lambda\}$. Then (C.21) proves (C.16), as desired. □

Step 3. Since, by Step 1, the pair $\{\tilde{u}_\lambda, \tilde{y}_\lambda\}$ in (C.15b) is a competitive pair for OCP #2, while $\{\mu_\lambda^0, z_\lambda^0\}$ is the optimal pair, then (C.16) yields at once

$$
\begin{aligned}
\text{(C.22)} \quad (Ry_0, y_0)_H &= J_\lambda\left(\tilde{u}_\lambda(\,\cdot\,, y_0), \tilde{y}_\lambda(\,\cdot\,; y_0)\right) \geq J_\lambda\left(\mu_\lambda^0(\,\cdot\,; y_0), z_\lambda^0(\,\cdot\,; y_0)\right) \\
&= J_\lambda^0(y_0) = (R_\lambda y_0, y_0)_H, \ \forall \ y_0 \in W, \quad \text{or } R \geq R_\lambda,
\end{aligned}
$$

recalling (C.7) in the last step. Thus, $R \geq R_\lambda$.

Step 4. It remains to show that

$$\text{(C.23)} \qquad (R_\lambda y_0, y_0)_H \geq (Ry_0, y_0)_H, \ \forall \ y_0 \in W, \qquad \text{or } R_\lambda \geq R.$$

To this end, we shall establish the following revealing identity

$$
\begin{aligned}
\text{(C.24)} \quad (R_\lambda y_0, y_0)_H &= \int_0^\infty \left[|Qe^{\lambda t}z_\lambda^0(t; y_0)|_H^2 + |e^{\lambda t}\mu_\lambda^0(t; y_0)|_U^2\right] dt \\
&\quad + 2\lambda \int_0^\infty \left((R - R_\lambda)e^{\lambda t}z_\lambda^0(t; y_0), e^{\lambda t}z_\lambda^0(t; y_0)\right)_H dt.
\end{aligned}
$$

Once (C.24) is established, we can readily prove (C.23). In fact, since $\lambda > 0$ and $R \geq R_\lambda$ by Step 3, the last term in identity (C.24) is non-negative. Moreover, by Step 1, the pair $\{e^{\lambda t}z_\lambda^0(t; y_0), e^{\lambda t}\mu_\lambda^0(t; y_0)\}$ is a competitive pair for OCP #1, while $\{u^0(t; y_0), y^0(t; y_0)\}$ is the optimal pair. Thus, all these considerations applied to identity (C.24) yield

$$
\begin{aligned}
\text{(C.25)} \quad (R_\lambda y_0, y_0)_H &\geq \int_0^\infty \left[|Qe^{\lambda t}z_\lambda^0(t; y_0)|_H^2 + |e^{\lambda t}\mu_\lambda^0(t; y_0)|_U^2\right] dt \\
\text{(C.26)} \quad &= J\left(e^{\lambda \cdot}\mu_\lambda^0(\,\cdot\,; y_0), e^{\lambda \cdot}z_\lambda^0(\,\cdot\,; y_0)\right) \\
&\geq J\left(u^0(\,\cdot\,; y_0), y^0(\,\cdot\,; y_0)\right) \\
&= J^0(y_0) = (Ry_0, y_0)_H, \ \forall \ y_0 \in W,
\end{aligned}
$$

and (C.23) is then obtained.

Step 5. It remains to prove identity (C.24). To this end, we begin by computing, by virtue of (C.9) and interchanging the order of integration

$$
\begin{aligned}
\text{(C.27)} \quad \int_0^\infty &\left(R_\lambda e^{\lambda t}z_\lambda^0(t; y_0), e^{\lambda t}z_\lambda^0(t; y_0)\right)_H dt \\
&= \int_0^\infty e^{2\lambda t} \int_t^\infty \left\{([Q^*Q + 2\lambda R]z_\lambda^0(\tau; y_0), z_\lambda^0(\tau; y_0))_H \right. \\
&\qquad\qquad \left. + |\mu_\lambda^0(\tau; y_0)|_U^2\right\} d\tau \, dt \\
\text{(C.28)} \quad &= \int_0^\infty \left\{([Q^*Q + 2\lambda R]z_\lambda^0(\tau; y_0), z_\lambda^0(\tau; y_0))_H \right. \\
&\qquad\qquad \left. + |\mu_\lambda^0(\tau; y_0)|_U^2\right\} \left(\int_0^\tau e^{2\lambda t} dt\right) d\tau.
\end{aligned}
$$

Using $\int_0^\tau e^{2\lambda t} dt = (e^{2\lambda \tau} - 1)/2\lambda$ in (C.28), and recalling (C.7) for R_λ, we obtain

$$(C.29) \quad \int_0^\infty \left(R_\lambda e^{\lambda t} z_\lambda^0(t; y_0), e^{\lambda t} z_\lambda^0(t; y_0)\right)_H dt$$

$$= \frac{1}{2\lambda} \int_0^\infty \left\{ ([Q^*Q + 2\lambda R]e^{\lambda \tau} z_\lambda^0(\tau; y_0), e^{\lambda \tau} z_\lambda^0(\tau; y_0))_H \right.$$

$$\left. + |e^{\lambda \tau} \mu_\lambda^0(\tau; y_0)|_U^2 \right\} d\tau$$

$$- \frac{1}{2\lambda} \int_0^\infty \left\{ ([Q^*Q + 2\lambda R] z_\lambda^0(\tau; y_0), z_\lambda^0(\tau; y_0))_H + |\mu_\lambda^0(\tau; y_0)|_U^2 \right\} d\tau$$

$$(C.30) \quad = \frac{1}{2\lambda} \int_0^\infty \left\{ |Q e^{\lambda \tau} z_\lambda^0(\tau; y_0)|_H^2 + 2\lambda \left(R e^{\lambda \tau} z_\lambda^0(\tau; y_0), e^{\lambda \tau} z_\lambda^0(\tau; y_0) \right)_H \right.$$

$$\left. + |e^{\lambda \tau} \mu_\lambda^0(\tau; y_0)|_U^2 \right\} d\tau - \frac{1}{2\lambda} (R_\lambda y_0, y_0)_H,$$

where, in the last step, we have used that the second integral term on the RHS of (C.29) is precisely $(R_\lambda y_0, y_0)_H$ by (C.7). Thus, (C.30) readily yields (C.24), as desired. We have thus proved that $(Ry_0, y_0)_H = (R_\lambda y_0, y_0)_H$, $\forall \, y_0 \in W$, i.e., $R_\lambda = R$, and part (ii) is proved.

(i) Having established $R_\lambda = R$, we return to (C.24) and obtain by recalling (C.7) that

$$(C.31) \quad (R_\lambda y_0, y_0)_H = \int_0^\infty \left[|Q e^{\lambda t} z_\lambda^0(t; y_0)|_H^2 + |e^{\lambda t} \mu_\lambda^0(t; y_0)|_U^2 \right] dt = J_\lambda^0(y_0).$$

But $\{e^{\lambda t} z_\lambda^0(t; y_0), e^{\lambda t} \mu_\lambda^0(t; y_0)\}$ is a competitive pair of OCP #1 by Step 1. Then, by uniqueness of the optimal pair $\{u^0(t; y_0), y^0(t; y_0)\}$ of OCP #1, this implies that

$$u^0(t; y_0) = e^{\lambda t} \mu_\lambda^0(t; y_0); \quad y^0(t; y_0) = e^{\lambda t} z_\lambda^0(t; y_0),$$

and (C.10) is established. Part (i) is likewise proved.

Parts (iii) and (iv) follow as counterparts of (4.3.21), (4.3.13), respectively.

Second proof of the key Steps 4-5. *Step (a)*. The key new step of the present proof is contained in the next result which gives a version of the "maximum principle."

PROPOSITION C.2. *Let $\{u(t), y(t) \equiv y(t; y_0)\}$ be a compatible pair for the dynamics (C.1): $y_t(t) = \mathcal{A} y(t) + \mathcal{B} u(t)$, $y(0) = y_0$. Assume further that $u \in L^2(0, \infty; U)$ and that $y(t)$ is differentiable on W. Then, the following inequality holds true for all $t \geq 0$:*

$$(C.32) \quad \frac{d}{dt}(Ry(t), y(t))_H \geq -[|Qy(t)|_H^2 + |u(t)|_U^2], \quad \forall \, t \geq 0.$$

PROOF OF PROPOSITION C.2. We shall prove separately that the right derivative $\frac{d^+}{dt}$ and the left derivative $\frac{d^-}{dt}$ satisfy inequality (C.32). We first recall R from

(C.3):

(C.33) $(Ry(t), y(t))_H = J(u^0(\cdot, 0; y(t)), y^0(\cdot, 0; y(t)))$

$$= \int_0^\infty [\|Qy^0(\tau, 0; y(t))\|_H^2 + |u^0(\tau, 0; y(t))|_U^2] d\tau$$

$$= \int_t^\infty [\|Qy^0(\tau - t, 0; y(t))\|_H^2 + |u^0(\tau - t, 0; y(t))|_U^2] d\tau$$

$$= \int_t^\infty [\|Qy^0(\tau, t; y(t))\|_H^2 + |u^0(\tau, t; y(t))|_U^2] d\tau,$$

where in the last step we have used the "optimality principle": the semigroup properties (4.2.5), (4.2.6) of the optimal pair. Similarly to (C.33), for $\Delta t > 0$, we obtain

(C.34) $(Ry(t + \Delta t), y(t + \Delta t))_H = \int_{t+\Delta t}^\infty [\|Qy^0(\tau, t + \Delta t; y(t + \Delta t))\|_H^2$

$$+ |u^0(\tau, t + \Delta t; y(t + \Delta t))|_U^2] d\tau;$$

(C.35) $(Ry(t - \Delta t), y(t - \Delta t))_H = \int_{t-\Delta t}^\infty [\|Qy^0(\tau, t - \Delta t; y(t - \Delta t))\|_H^2$

$$+ |u^0(\tau, t - \Delta t; y(t - \Delta t))|_U^2] d\tau.$$

Proof of (C.32) for $\frac{d^+}{dt}$. In order to establish inequality (C.32) for the right derivative $\frac{d^+}{dt}(Ry(t), y(t))_H$, $t \geq 0$, we construct the following 'comparison' function:

(C.36) $y^*(\tau) = \begin{cases} y(\tau); \\ y^0(\tau, t + \Delta t; y(t + \Delta t)); \end{cases}$

$u^*(\tau) = \begin{cases} u(\tau), & \text{for } \tau \in [t, t + \Delta t]; \\ u^0(\tau, t + \Delta t; y(t + \Delta t)), & \text{for } \tau > t + \Delta t. \end{cases}$

In particular, for $\tau = t$, we have: $y^*(t) = y(t)$, $u^*(t) = u(t)$; and so $\{u^*(t), y^*(t)\}$ is a compatible pair for Eqn. (C.1). Accordingly, from the optimal cost (C.33), we obtain

(C.37) $$(Ry(t), y(t))_H \leq \int_t^\infty [\|Qy^*(\tau)\|_H^2 + |u^*(\tau)|_U^2] d\tau.$$

On the other hand, by the definition of $\{u^*, y^*\}$ for $\tau > t + \Delta t$ in (C.36b), used in the optimal cost (C.34), we obtain

(C.38) $$(Ry(t + \Delta t), y(t + \Delta t))_H = \int_{t+\Delta t}^\infty [\|Qy^*(\tau)\|_H^2 + |u^*(\tau)|_U^2] d\tau.$$

Hence, by subtracting the LHS of (C.38) and (C.37), we obtain

(C.39) $\quad (Ry(t+\Delta t), y(t+\Delta t))_H - (Ry(t), y(t))_H$

$$\geq \int_{t+\Delta t}^{\infty} [|Qy^*(\tau)|_H^2 + |u^*(\tau)|_U^2] d\tau$$

$$- \int_t^{\infty} [|Qy^*(\tau)|_H^2 + |u^*(\tau)|_U^2] d\tau$$

(C.40) $\quad = -\int_t^{t+\Delta t} [|Qy^*(\tau)|_H^2 + |u^*(\tau)|_U^2] d\tau$

(C.41) $\quad$ (by (C.36a)) $\quad = -\int_t^{t+\Delta t} [|Qy(\tau)|_H^2 + |u(\tau)|_U^2] d\tau,$

where in the last step we have recalled (C.36a) on $[t, t+\Delta t]$. Finally, dividing inequality (C.41) across by $\Delta t > 0$, and taking the limit $\Delta t \searrow 0$ yields inequality (C.32) for the right derivative $\frac{d^+}{dt}(Ry(t), y(t))_H$, $t \geq 0$, as desired.

Proof of (C.32) for $\frac{d^-}{dt}$. In order to establish inequality (C.32) for the left derivative $\frac{d^-}{dt}(Ry(t), y(t))_H$, $t > 0$, we introduce a different 'comparison' function, still with $\Delta t > 0$:

(C.42) $\quad y^*(\tau) = \begin{cases} y(\tau); \\ y^0(\tau, t; y(t)); \end{cases} \quad u^*(\tau) = \begin{cases} u(\tau), & \tau \in [t-\Delta t, t]; \\ u^0(\tau, t; y(t)), & \tau > t. \end{cases}$

In particular, for $\tau = t$, we have: $y^*(t) = y(t)$, $u^*(t) = u(t)$; and so $\{u^*(t), y^*(t)\}$ is a compatible pair for Eqn. (C.1). Accordingly, from the optimal cost (C.33), we obtain

(C.43) $\quad (Ry(t-\Delta t), y(t-\Delta t))_H \leq \int_{t-\Delta t}^{\infty} [|Qy^*(\tau)|_H^2 + |u^*(\tau)|_U^2] d\tau.$

On the other hand, by the definition of $\{u^*, y^*\}$ for $\tau > t$ in (C.42b), used in the optimal cost (C.33), we obtain

(C.44) $\quad (Ry(t), y(t))_H = \int_t^{\infty} [|Qy^*(\tau)|_H^2 + |u^*(\tau)|_U^2] d\tau.$

Hence, by subtracting the LHS of (C.43) and (C.44), we obtain

(C.45) $\quad (Ry(t), y(t))_H - (Ry(t-\Delta t), y(t-\Delta t))_H$

$$\geq \int_t^{\infty} [|Qy^*(\tau)|_H^2 + |u^*(\tau)|_U^2] d\tau$$

$$- \int_{t-\Delta t}^{\infty} [|Qy^*(\tau)|_H^2 + |u^*(\tau)|_U^2] d\tau$$

(C.46) $\quad = -\int_t^{t-\Delta t} [|Qy^*(\tau)|_H^2 + |u^*(\tau)|_U^2] d\tau$

(C.47) $\quad$ (by (C.42a)) $\quad = -\int_{t-\Delta t}^{t} [|Qy(\tau)|_H^2 + |u(\tau)|_U^2] d\tau,$

C. EQUIVALENCE BETWEEN UNSTABLE AND STABLE VERSIONS   121

where in the last step we have recalled (C.42a) on $[t - \Delta t, t]$. Finally, dividing inequality (C.47) across by $\Delta t > 0$, and taking the limit $\Delta t \searrow 0$ yields inequality (C.32) for the left derivative $\frac{d^-}{dt}(Ry(t), y(t))_H$, $t > 0$, as desired. Thus Proposition C.2 is proved. □

Step (b). In this step, by using Proposition C.2, we shall establish (C.23): $R_\lambda \geq R$, or $J_\lambda^0(y_0) = J_\lambda(u_\lambda^0, y_\lambda^0) \geq J^0(y_0) = J(u^0, y^0)$. In fact, using the notation: $u(t) \equiv u_\lambda^0(t)e^{\lambda t}$, $y(t) \equiv y^0(t)e^{\lambda t}$, so that $\{u(t), y(t)\}$ is a compatible pair for (C.1): $y_t(t) = \mathcal{A}y(t) + \mathcal{B}u(t)$ by (C.15a), we compute:

$$(C.48) \quad J_\lambda(u_\lambda^0(t), y_\lambda^0(t))$$

$$= \int_0^\infty [|Qy_\lambda^0(t)|_H^2 + |u_\lambda^0(t)|_U^2 + 2\lambda(Ry_\lambda^0(t), y_\lambda^0(t))_H]dt$$

$$(C.49) \quad = \int_0^\infty e^{-2\lambda t}[|Qy(t)|_H^2 + |u(t)|_U^2]dt$$

$$+ \int_0^\infty 2\lambda e^{-2\lambda t}(Ry(t), y(t))_H dt.$$

Integration by parts on the last term of (C.49) yields

$$(C.50) \quad -\int_0^\infty \frac{de^{-2\lambda t}}{dt}(Ry(t), y(t))_H dt$$

$$= (Ry_0, y_0)_\Omega + \int_0^\infty e^{-2\lambda t}\frac{d}{dt}(Ry(t), y(t))_\Omega dt,$$

since

$$(C.51) \quad e^{-2\lambda t}(Ry(t), y(t))_H = (Ry_\lambda^0(t), y_\lambda^0(t))_H \to 0, \text{ as } t \to \infty.$$

To see (C.51), we note that, since $\lambda > 0$, then by (C.5c):

$$(C.52) \quad \int_0^\infty |y_\lambda^0(t; y_0)|_W^2 dt \leq J_\lambda(u_\lambda^0, y_\lambda^0) < \infty, \text{ where}$$

$$|y_\lambda^0(t; y_0)|_W^2 \sim (Ry^0(t; y_0), y_\lambda^0(t; y_0))_H$$

by (4.1.13b). Moreover, $y_\lambda^0(t; y_0) \equiv S_\lambda(t)y_0$, where $S_\lambda(t)$ is a s.c. semigroup on W [same reasons as for Proposition 4.2.2, this time for the OCP #2]. Then Datko's result [**P.1**, p. 116] applied on (C.52) yields that $|y_\lambda^0(t; y_0)|_W$ decays exponentially, and (C.51) is established with an exponential rate. Next, we specialize to initial conditions $y_0 \in \mathcal{D}(A_R)$ so that $y(t) = y(t; y_0)$ is differentiable on W. In this case, we can invoke (C.32) of Proposition C.2 on (C.50), and use the resulting inequality on the RHS of (C.49), thereby obtaining

$$(C.53) \quad J_\lambda(u_\lambda^0(t), y_\lambda^0(t)) \geq \int_0^\infty e^{-2\lambda t}[|Qy(t)|_H^2 + |u(t)|_U^2]dt + (Ry_0, y_0)_H$$

$$- \int_0^\infty e^{-2\lambda t}[|Qy(t)|_H^2 + |u(t)|_U^2]dt = (Ry_0, y_0)_H = J(u^0, y^0),$$

initially for $y_0 \in \mathcal{D}(A_R)$, and then to all $y_0 \in W$ by density. Then (C.53) proves our claim: $R_\lambda \geq R$.

APPENDIX D

Proof that $FS(\,\cdot\,) \in \mathcal{L}(W; L^2(0,\infty;(L^2(\Gamma))^d)$

We recall that in the proof of Lemma 5.3, the argument leading to Eqn. (5.15) and line immediately below yields that inequality (5.7) holds true for all data $\{z_0, f\}$ defined by (5.10). In particular, for $f = 0$ and $x \in \mathcal{D}(A_R^2)$, this result yields the first of the following two results:

(D.0) $\begin{cases} \int_0^\infty |FS(t)x|_U^2 dt \leq C|x|_W^2, & x \in \mathcal{D}(A_R^2), \\ \text{for a positive constant } C > 0 \text{ independent of } x, \\ FS(t)x = D^*\mathcal{A}^*RS(t)x = \nu_0 \dfrac{\partial}{\partial \nu} RS(t)x \in (H^{1+\epsilon}(\Gamma))^d, & x \in \mathcal{D}(A_R^2), \end{cases}$

while the second such result is precisely (4.4.20).

We recall from [(4.3.30) and hence from] (4.4.16) that, by definition, $Fx \equiv D^*\mathcal{A}^*Rx \in (H^{1+\epsilon}(\Gamma))^d$, $\forall\, x \in \mathcal{D}(A_R^2)$. Since $\mathcal{D}(A_R^2)$ is dense in W, inequality (D.0) says that the operator $FS(\,\cdot\,)$, originally defined and continuous from $\mathcal{D}(A_R^2)$ into $L^2(0,\infty;(L^2(\Gamma))^d)$, possesses a continuous extension from W to $L^2(0,\infty;(L^2(\Gamma))^d)$. The present Appendix shows more: that such an extension continues to be $FS(\,\cdot\,)$, as anticipated in Remark 4.4.1.

PROPOSITION D.1. *Let $x \in W$ so that there exists a sequence $\{x_n\}$, $x_n \in \mathcal{D}(A_R^2)$, such that $x_n \to x$ in W and (D.0) holds true with x replaced by x_n. Then, in fact, the following convergence holds true, as $n \to \infty$:*

(D.1) $\quad FS(\,\cdot\,)x_n \equiv D^*\mathcal{A}^*RS(\,\cdot\,)x_n = \nu_0 \dfrac{\partial}{\partial \nu} RS(t)x_n$

$\qquad \longrightarrow FS(\,\cdot\,)x \equiv D^*\mathcal{A}^*RS(\,\cdot\,)x = \nu_0 \dfrac{\partial}{\partial \nu} RS(\,\cdot\,)x$

in $L^2(0,\infty;(L^2(\Gamma))^d)$, $\dfrac{\partial RS(t)x}{\partial \nu} \cdot \nu = 0$ on Γ.

PROOF. Recalling (4.4.16), (4.4.20) for F as well as (4.4.12) for R, we see that if $x_n \in \mathcal{D}(A_R^2)$, then the following identity holds true via the semigroup property of $S(\,\cdot\,)$:

(D.2) $\quad FS(t)x_n \equiv D^*\mathcal{A}^*RS(t)x_n = D^*\mathcal{A}^* \int_0^\infty e^{\mathcal{A}^*\tau} Q^*QS(\tau+t)x_n d\tau.$

Let $x \in W$ and $\mathcal{D}(A_R^2) \ni x_n \to x$ in W. To establish the desired conclusion (D.1), it suffices to show that, as $n \to \infty$, then

(D.3) $\quad \int_0^\infty (FS(t)x_n, \varphi(t))_U dt \to \int_0^\infty (FS(t)x, \varphi(t))_U dt,$

for all $\varphi \in C_0^2(0,\infty;(H^{1+\epsilon}(\Gamma))^d)$, $\varphi \cdot \nu \equiv 0$ on Γ, since $C_0^2(0,\infty;(H^{1+\epsilon}(\Gamma))^d)$ is dense in $L^2(0,\infty(L^2(\Gamma))^d)$. Here $FS(t)x$ is defined by (D.2) with $x_n \in \mathcal{D}(A_R^2)$ replaced by $x \in W$. We compute via (D.2):

$$(D.4) \quad \int_0^\infty (FS(t)x_n, \varphi(t))_U \, dt = \int_0^\infty \int_0^\infty (QS(\tau+t)x_n, Q\mathcal{A}e^{\mathcal{A}\tau}D\varphi(t))_H \, d\tau \, dt.$$

Henceforth, we shall interchange the order of integration, as needed, with no change of the limits of integration. Changing the order of integration on the RHS of (D.4) and integrating by parts twice in t, we obtain, after recalling that φ has compact support on $(0, \infty)$:

$$(D.5) \quad \int_0^\infty (FS(t)x_n), \varphi(t))_U \, dt$$

$$= \int_0^\infty \int_0^\infty (QS(\tau+t)x_n, Q\mathcal{A}e^{\mathcal{A}\tau}D\varphi(t))_H \, dt \, d\tau$$

(D.6) (by (4.2.9))

$$= \int_0^\infty \int_0^\infty (QA_R^{-2}S(\tau+t)x_n, Q\mathcal{A}e^{\mathcal{A}\tau}D\varphi_{tt}(t))_H \, dt \, d\tau$$

$$(D.7) \quad = \int_0^\infty \int_0^\infty (QA_R^{-2}S(\tau+t)x_n, Q\mathcal{A}e^{\mathcal{A}\tau}D\varphi_{tt}(t))_H \, d\tau \, dt$$

(D.8) (by (4.2.9))

$$= \int_0^\infty \left[(QA_R^{-2}S(\tau+t)x_n, Qe^{\mathcal{A}\tau}D\varphi_{tt}(t))_H\right]_{\tau=0}^{\tau=\infty} dt$$

$$- \int_0^\infty \int_0^\infty (QA_R^{-1}S(\tau+t)x_n, Qe^{\mathcal{A}\tau}D\varphi_{tt}(t))_H \, d\tau \, dt$$

(D.9) (by (4.2.11))

$$= -\int_0^\infty (QA_R^{-2}S(t)x_n, QD\varphi_{tt}(t))_H \, dt$$

$$- \int_0^\infty \int_0^\infty (QA_R^{-1}S(\tau+t)x_n, Qe^{\mathcal{A}\tau}D\varphi_{tt}(t))_H \, d\tau \, dt$$

$$\longrightarrow \text{(converging, as } n \to \infty, \text{ to)}$$

$$(D.10) \quad -\int_0^\infty (QA_R^{-2}S(t)x, QD\varphi_{tt}(t))_H \, dt$$

$$- \int_0^\infty \int_0^\infty (QA_R^{-1}S(\tau+t)x, Qe^{\mathcal{A}\tau}D\varphi_{tt}(t))_H \, d\tau \, dt$$

(D.11) (by (D.5))

$$= \int_0^\infty \int_0^\infty (QS(\tau+t)x, Q\mathcal{A}e^{\mathcal{A}\tau}D\varphi(t))_H \, dt \, d\tau$$

(D.12) (by (D.5))

$$= \int_0^\infty (FS(t)x, \varphi(t))_U \, dt.$$

We now justify the steps. In the double integration by parts in t from (D.5) to (D.6), we have used (4.2.9) as well as $\varphi(0) = \varphi_t(0) = \varphi(\infty) = \varphi_t(\infty) = 0$, by assumption. In going from (D.6) to (D.7) we have interchanged the order of integration, so that a subsequent integration by parts in τ leads to (D.8). In going from (D.8) to (D.9), we have invoked estimate (4.2.11)

(D.13) $\quad |QS(\tau + t)A_R^{-2}x_n|_H \leq Ce^{-a(\tau+t)}|A_R^{-1}x_n|_W \to 0$ as $\tau \to \infty$.

We note that (D.9) is well defined. This is so for two reasons. First, $QD\varphi_{tt}(t) \in H$, since $\varphi_{tt}(t) \in (H^{1+\epsilon}(\Gamma))^d$ by assumption, hence $D\varphi_{tt}(t) \in (H^{\frac{3}{2}+\epsilon}(\Omega))^d \cap H$ by (3.1.3d) with $\varphi_{tt}(t) \cdot \nu \equiv 0$ on Γ; finally we invoke (4.1.1) on Q.

We note that (D.9) is well-defined. This is so for two reasons. The first reason is that

(D.14) $\quad\quad\quad QD\varphi_{tt}(t) \in H$ and, moreover, $\varphi_{tt} \in C_0(0,\infty)$.

In fact, we have $\varphi_{tt}(t) \in (H^{1+\epsilon}(\Gamma))^d$ by assumption, hence $D\varphi_{tt}(t) \in (H^{\frac{3}{2}+\epsilon}(\Omega))^d \cap H$ by (3.1.3d) with $\varphi_{tt}(t) \cdot \nu = 0$ on Γ; finally, (4.1.1) on Q yields the LHS on (D.14).

The second version is that

$$\text{(D.15)} \quad |Qe^{\mathcal{A}\tau}D\varphi_{tt}(t)|_H = \left|Q(-\mathcal{A})^{-(\frac{3}{4}+\frac{\epsilon}{2})}(-\mathcal{A})^{\frac{1}{2}+\epsilon}e^{\mathcal{A}\tau}(-\mathcal{A})^{\frac{1}{4}-\frac{\epsilon}{2}}D\varphi_{tt}(t)\right|_H$$
$$\leq \frac{Ce^{-\delta\tau}}{\tau^{\frac{1}{2}+\epsilon}}|\varphi_{tt}(t)|_U,$$

where we have invoked (4.3.2b); (4.3.0) and analyticity of $e^{\mathcal{A}\tau}$; as well as (3.1.3d) along with $\varphi_{tt}(t) \cdot \nu = 0$ to claim that $(-\mathcal{A})^{\frac{1}{4}-\frac{\epsilon}{2}}D\varphi_{tt}(t) \in H$. Thus, (D.15) follows.

The convergence of (D.9) to (D.10) as $n \to \infty$ is plain. Then, (D.10) leads to (D.11) and (D.12) by tracing backward for "x" the steps from (D.5) to (D.10) for "x_n." Each such term with $x_n \in \mathcal{D}(A_R^2)$ replaced by $x \in W$ is likewise well-defined: in other words, the steps from (D.5) to (D.10) do not depend on x_n being in $\mathcal{D}(A_R^2)$ and are equally valid for $x \in W$ only.

In conclusion, (D.5) $\to$ (D.12) and Proposition D.1 is established. $\square$

PROPOSITION D.2. *Let $f \in L^1(0,\infty;W)$ so that there exists a sequence $\{f_n\}$, $f_n \in L^1(0,\infty;\mathcal{D}(A_R^2))$ such that $f_n \to f$ in $L^1(0,\infty;W)$ and, by (4.4.16),*

$$\text{(D.16)} \quad F\int_0^t S(t-\tau)f_n(\tau)d\tau = D^*\mathcal{A}^*R\int_0^t S(t-\tau)f_n(\tau)d\tau$$
$$\in C([0,T];(H^{1+\epsilon}(\Gamma))^d) \cap L^2(0,\infty;(H^{1+\epsilon}(\Gamma))^d),$$

recalling (4.3.33). Then, in fact, the following convergence holds true, as $n \to \infty$, where $U = \{u \in (L^2(\Gamma))^d, u \cdot \nu$ on $\Gamma\}$ as in (4.1.1a):

$$\text{(D.17)} \quad F\int_0^t S(t-\tau)f_n(\tau)d\tau \equiv D^*\mathcal{A}^*R\int_0^t S(t-\tau)f_n(\tau)d\tau$$
$$\longrightarrow F\int_0^t S(t-\tau)f(\tau)d\tau$$
$$\equiv D^*\mathcal{A}^*R\int_0^t S(t-\tau)f(\tau)d\tau \text{ in } L^2(0,\infty;U).$$

PROOF. Let $\varphi(t) \in L^2(0,\infty;U)$. Then, interchanging the order of integration and invoking Proposition D.1, we obtain

(D.18) $$\int_0^\infty \left(\int_0^t FS(t-\tau)f_n(\tau)d\tau, \varphi(t) \right)_U dt$$
$$= \int_0^\infty \int_t^\infty (FS(t-\tau)f_n(\tau), \varphi(t))_U dt\, d\tau$$

(D.19) $$\longrightarrow \int_0^\infty \int_t^\infty (FS(t-\tau)f(\tau), \varphi(t))_U dt\, d\tau$$
$$= \int_0^\infty \left(\int_0^t FS(t-\tau)f(\tau)d\tau, \varphi(t) \right)_U d\tau\, dt,$$

as desired, and (D.17) is established. □

Bibliography

[A.1]   J. H. Albert, Genericity of simple eigenvalues for elliptic pde's, *Proc. AMS* 48(2) (1975), 413–418.

[Au.1] G. Auchmuty, Orthogonal decomposition and bases for 3-D vector fields, *Num. Funct. Anal. Optimiz.* 15 (1994), 445–488.

[Bal.1] A. V. Balakrishnan, *Applied Functional Analysis*, Second Edition, Springer-Verlag, 1981.

[B.1]   V. Barbu, Feedback stabilization of Navier-Stokes equations, *ESAIM COCV* 9 (2003), 197–206.

[B-T.1] V. Barbu, R. Triggiani, Internal stabilization of Navier-Stokes equations with finite-dimensional controllers, *Indiana University Mathematics Journal*, 53(5) (2004), 1443–1494.

[Ce.1] M. Cessenat, *Mathematical Methods in Electromagnetism*, World Scientific, 1996.

[C-F.1] P. Constantin, C. Foias, *Navier-Stokes Equations*, University of Chicago Press, Chicago, London, 1989.

[D-S.1] N. Dunford and J. T. Schwartz, *Linear Operators*, Part III: *Spectral Operators*, Wiley-Interscience, 1971.

[F-L-T.1] F. Flandoli, I. Lasiecka, R. Triggiani, Algebraic Riccati equations with non-smoothing observation arising in hyperbolic and Euler-Bernoulli boundary control problems, *Ann. Math. Pura e Appl.* CLIII (1988), 307–382.

[F-T.1] C. Foias, R. Teman, Remarques sur les equations de Navier-Stokes stationaires el les phenomenes successifs de bifurcation, *Annali Scuola Normale Sup.*, Pisa 1 (1978), 26–53.

[F-I.1] A. V. Fursikov, O. A. Imanuvilov, Local exact boundary controllability of the Navier-Stokes system, *Contemp. Math.* 209 (1997), 115–129.

[F-I.2] A. V. Fursikov, O. A. Imanuvilov, Exact controllability of the Navier-Stokes and Boussinesq equations, *Russian Math. Surveys* 54 (1999), 565–618.

[F.1]   A. V. Fursikov, Real processes to the 3D Navier-Stokes systems, and its feedback stabilization from the boundary, in *Amer. Math. Soc. Translations*, Series 2, Vol. 206, *Partial Differential Equations*, Mark Vishik's Seminar; M. S. Agranovich, M. A. Shubin, editors (2002), pp. 95–123.

[F.2]   A. V. Fursikov, Feedback stabilization for the 2D Oseen equations: Additional remarks, Proc. 8th Conference on Control of Distributed Parameter Systems, *Int. Ser. Numerical Math.*, Birkhäuser-Verlag 143 (2002), 169–187.

[F.3]   A. V. Fursikov, Stabilization for the 3D Navier-Stokes system by feedback boundary control, *Discrete and Contin. Dynam. Systems* 10(1&2) (2004), 289–314.

[G.1]   G. Galdi, *An Introduction to the Mathematical Theory of Navier-Stokes Equations*, Vol. 1, Springer, 1998.

[G-K.1] I. C. Goberg and M. G. Krein, Introduction to the Theory of Linear Non-selfadjoint Operators, Vol. 18, *Translations of Mathematical Monographs*, Amer. Math. Soc., 1969.

[K.1]   T. Kato, *Perturbation Theory of Linear Operators*, Springer-Verlag, New York, Berlin, 1966.

[K.2]   S. Kesavan, *Topics in Functional Analysis and Applications*, John Wiley & Sons, 1989.

[L.1]   O. A. Ladyzhenskaya, *The Boundary Value Problems of Mathematical Physics*, Springer-Verlag, 1985.

[L-T.1] I. Lasiecka and R. Triggiani, *Differential and Algebraic Riccati Equations with Applications to Boundary/Point Control Problems*, LNCIS, vol. 164, 160 pp., Springer Verlag, 1991.

[L-T.2]	I. Lasiecka and R. Triggiani, *Control Theory for Partial Differential Equations: Continuous and Approximation Theory*, Vol. 1 (680 pp.), Vol. 2 (422 pp.), Encyclopedia of Mathematics and its Applications, Cambridge University Press, 2000.
[L-M.1]	J. L. Lions and E. Magenes, *Non-homogeneous Boundary Value Problems and Applications*, Vol. I (1970), Springer Verlag, Vol. 2 (1970).
[M.1]	A. M. Micheletti, Perturbazione dello spettro dell' operatore di Laplace in relazione ad una variazione del campo, *Ann. Scuola Normal Super.*, Pisa 26(3) (1972), 151–169.
[M-M.1]	R. Miller and A. N. Michel, *Ordinary Differential Equations*, AP 1982.
[M-S.1]	V. G. Maz'ya and T. O. Shaposhnikova, *Theory of Multipliers in Spaces of Differentiable Functions*, Pitman Advanced Publishing Program, 1985.
[P.1]	A. Pazy, *Semigroups of Linear Operators and Applications to Partial Differential Equations*, Springer-Verlag, 1983.
[R-S.1]	T. Runst and W. Sickel, *Sobolev Spaces of Fractional Order, Nemytskij Operators and Nonlinear Partial Differential Equations*, Walter de Gruyter–Berlin–New York, 1996.
[T-L.1]	A. Taylor and D. C. Lay, *Introduction to Functional Analysis*, 2nd edition, John Wiley, 1980.
[Te.1]	R. Temam, *Navier-Stokes Equations*, Studies in Mathematics and its Applications, Vol. 2, North-Holland Publishing Co., Amsterdam, 1979.
[Te.2]	R. Temam, *Navier-Stokes Equations and Nonlinear Functional Analysis*, SIAM, Philadelphia, 1983.
[Tr.1]	H. Triebel, *Interpolation Theory, Function Spaces, Differential Operators*, North Holland, Amsterdam, 1978.
[T.1]	R. Triggiani, On the stabilizability problem in Banach spaces, *J. Math. Anal. Appl.* (1975), 383–403.
[T.2]	R. Triggiani, Boundary feedback stabilizability of parabolic equations, *Appl. Math. Optimiz.* 6 (1980), 201–220.
[T.3]	R. Triggiani, On the relationship between the optimal quadratic cost problems in an infinite horizon, and on a finite horizon with final time penalization: The abstract hyperbolic case, in *Optimal Control of Differential Equations*, edited by N. Pavel, *Lecture Notes in Pure and Applied Mathematics*, vol. 160, Marcel Dekker 1994, pp. 311–323.
[U.1]	K. Uhlenbeck, Generic properties of eigenfunctions, *Amer. J. Math.* 98(4) (197), 1059–1078.
[W.1]	W. von Wahl, *The equations of Navier-Stokes and abstract parabolic equations*, Vieweg & Sohn, Braunschweig, 1985.

Editorial Information

To be published in the *Memoirs*, a paper must be correct, new, nontrivial, and significant. Further, it must be well written and of interest to a substantial number of mathematicians. Piecemeal results, such as an inconclusive step toward an unproved major theorem or a minor variation on a known result, are in general not acceptable for publication. Papers appearing in *Memoirs* are generally at least 80 and not more than 200 published pages in length. Papers less than 80 or more than 200 published pages require the approval of the Managing Editor of the Transactions/Memoirs Editorial Board.

As of January 31, 2006, the backlog for this journal was approximately 14 volumes. This estimate is the result of dividing the number of manuscripts for this journal in the Providence office that have not yet gone to the printer on the above date by the average number of monographs per volume over the previous twelve months, reduced by the number of volumes published in four months (the time necessary for preparing a volume for the printer). (There are 6 volumes per year, each containing at least 4 numbers.)

A Consent to Publish and Copyright Agreement is required before a paper will be published in the *Memoirs*. After a paper is accepted for publication, the Providence office will send a Consent to Publish and Copyright Agreement to all authors of the paper. By submitting a paper to the *Memoirs*, authors certify that the results have not been submitted to nor are they under consideration for publication by another journal, conference proceedings, or similar publication.

Information for Authors

Memoirs are printed from camera copy fully prepared by the author. This means that the finished book will look exactly like the copy submitted.

The paper must contain a *descriptive title* and an *abstract* that summarizes the article in language suitable for workers in the general field (algebra, analysis, etc.). The *descriptive title* should be short, but informative; useless or vague phrases such as "some remarks about" or "concerning" should be avoided. The *abstract* should be at least one complete sentence, and at most 300 words. Included with the footnotes to the paper should be the 2000 *Mathematics Subject Classification* representing the primary and secondary subjects of the article. The classifications are accessible from www.ams.org/msc/. The list of classifications is also available in print starting with the 1999 annual index of *Mathematical Reviews*. The Mathematics Subject Classification footnote may be followed by a list of *key words and phrases* describing the subject matter of the article and taken from it. Journal abbreviations used in bibliographies are listed in the latest *Mathematical Reviews* annual index. The series abbreviations are also accessible from www.ams.org/publications/. To help in preparing and verifying references, the AMS offers MR Lookup, a Reference Tool for Linking, at www.ams.org/mrlookup/. When the manuscript is submitted, authors should supply the editor with electronic addresses if available. These will be printed after the postal address at the end of the article.

Electronically prepared manuscripts. The AMS encourages electronically prepared manuscripts, with a strong preference for $\mathcal{AMS}$-LaTeX. To this end, the Society has prepared $\mathcal{AMS}$-LaTeX author packages for each AMS publication. Author packages include instructions for preparing electronic manuscripts, the *AMS Author Handbook*, samples, and a style file that generates the particular design specifications of that publication series. Though $\mathcal{AMS}$-LaTeX is the highly preferred format of TeX, author packages are also available in $\mathcal{AMS}$-TeX.

Authors may retrieve an author package from e-MATH starting from www.ams.org/tex/ or via FTP to ftp.ams.org (login as anonymous, enter username as password, and type cd pub/author-info). The *AMS Author Handbook* and the *Instruction Manual* are available in PDF format following the author packages link from www.ams.org/tex/. The author package can also be obtained free of charge by sending email to `tech-support@ams.org` (Internet) or from the Publication Division, American Mathematical Society, 201 Charles St., Providence, RI 02904-2294, USA. When requesting an author package, please specify $\mathcal{AMS}$-LaTeX or $\mathcal{AMS}$-TeX and the publication in which your paper will appear. Please be sure to include your complete mailing address.

Sending electronic files. After acceptance, the source file(s) should be sent to the Providence office (this includes any TeX source file, any graphics files, and the DVI or PostScript file).

Before sending the source file, be sure you have proofread your paper carefully. The files you send must be the EXACT files used to generate the proof copy that was accepted for publication. For all publications, authors are required to send a printed copy of their paper, which exactly matches the copy approved for publication, along with any graphics that will appear in the paper.

TeX files may be submitted by email, FTP, or on diskette. The DVI file(s) and PostScript files should be submitted only by FTP or on diskette unless they are encoded properly to submit through email. (DVI files are binary and PostScript files tend to be very large.)

Electronically prepared manuscripts can be sent via email to `pub-submit@ams.org` (Internet). The subject line of the message should include the publication code to identify it as a Memoir. TeX source files, DVI files, and PostScript files can be transferred over the Internet by FTP to the Internet node `e-math.ams.org` (130.44.1.100).

Electronic graphics. Comprehensive instructions on preparing graphics are available at `www.ams.org/jourhtml/graphics.html`. A few of the major requirements are given here.

Submit files for graphics as EPS (Encapsulated PostScript) files. This includes graphics originated via a graphics application as well as scanned photographs or other computer-generated images. If this is not possible, TIFF files are acceptable as long as they can be opened in Adobe Photoshop or Illustrator. No matter what method was used to produce the graphic, it is necessary to provide a paper copy to the AMS.

Authors using graphics packages for the creation of electronic art should also avoid the use of any lines thinner than 0.5 points in width. Many graphics packages allow the user to specify a "hairline" for a very thin line. Hairlines often look acceptable when proofed on a typical laser printer. However, when produced on a high-resolution laser imagesetter, hairlines become nearly invisible and will be lost entirely in the final printing process.

Screens should be set to values between 15% and 85%. Screens which fall outside of this range are too light or too dark to print correctly. Variations of screens within a graphic should be no less than 10%.

Inquiries. Any inquiries concerning a paper that has been accepted for publication should be sent directly to the Electronic Prepress Department, American Mathematical Society, 201 Charles St., Providence, RI 02904, USA.

Editors

This journal is designed particularly for long research papers, normally at least 80 pages in length, and groups of cognate papers in pure and applied mathematics. Papers intended for publication in the *Memoirs* should be addressed to one of the following editors. In principle the Memoirs welcomes electronic submissions, and some of the editors, those whose names appear below with an asterisk (*), have indicated that they prefer them. However, editors reserve the right to request hard copies after papers have been submitted electronically. Authors are advised to make preliminary email inquiries to editors about whether they are likely to be able to handle submissions in a particular electronic form.

*Algebra to ALEXANDER KLESHCHEV, Department of Mathematics, University of Oregon, Eugene, OR 97403-1222; email: ams@noether.uoregon.edu

Algebra and its application to MINA TEICHER, Emmy Noether Research Institute for Mathematics, Bar-Ilan University, Ramat-Gan 52900, Israel; email: teicher@macs.biu.ac.il

Algebraic geometry to DAN ABRAMOVICH, Department of Mathematics, Brown University, Box 1917, Providence, RI 02912; email: amsedit@math.brown.edu

*Algebraic number theory to V. KUMAR MURTY, Department of Mathematics, University of Toronto, 100 St. George Street, Toronto, ON M5S 1A1, Canada; email: murty@math.toronto.edu

*Algebraic topology to ALEJANDRO ADEM, Department of Mathematics, University of British Columbia, Room 121, 1984 Mathematics Road, Vancouver, British Columbia, Canada V6T 1Z2; email: adem@math.ubc.ca

Combinatorics to JOHN R. STEMBRIDGE, Department of Mathematics, University of Michigan, Ann Arbor, Michigan 48109-1109; email: FRS@umich.edu

Complex analysis and harmonic analysis to ALEXANDER NAGEL, Department of Mathematics, University of Wisconsin, 480 Lincoln Drive, Madison, WI 53706-1313; email: nagel@math.wisc.edu

*Differential geometry and global analysis to LISA C. JEFFREY, Department of Mathematics, University of Toronto, 100 St. George St., Toronto, ON Canada M5S 3G3; email: jeffrey@math.toronto.edu

Dynamical systems and ergodic theory to AMIE WILKINSON, Department of Mathematics, Northwestern University, 2033 Sheridan Road, Evanston, IL 60208-2730; email: wilkinso@math.northwestern.edu

*Functional analysis and operator algebras to MARIUS DADARLAT, Department of Mathematics, Purdue University, 150 N. University St., West Lafayette, IN 47907-2067; email: mdd@math.purdue.edu

*Geometric analysis to TOBIAS COLDING, Courant Institute, New York University, 251 Mercer St., New York, NY 10012; email: traneditor@cims.nyu.edu

*Geometric analysis to MLADEN BESTVINA, Department of Mathematics, University of Utah, 155 South 1400 East, JWB 233, Salt Lake City, Utah 84112-0090; email: bestvina@math.utah.edu

Harmonic analysis, representation theory, and Lie theory to ROBERT J. STANTON, Department of Mathematics, The Ohio State University, 231 West 18th Avenue, Columbus, OH 43210-1174; email: stanton@math.ohio-state.edu

*Logic to STEFFEN LEMPP, Department of Mathematics, University of Wisconsin, 480 Lincoln Drive, Madison, Wisconsin 53706-1388; email: lempp@math.wisc.edu

*Ordinary differential equations, and applied mathematics to PETER W. BATES, Department of Mathematics, Michigan State University, East Lansing, MI 48824-1027; email: bates@math.msu.edu

*Partial differential equations to GUSTAVO PONCE, Department of Mathematics, South Hall, Room 6607, University of California, Santa Barbara, CA 93106; email: ponce@math.ucsb.edu

*Probability and statistics to KRZYSZTOF BURDZY, Department of Mathematics, University of Washington, Box 354350, Seattle, Washington 98195-4350; email: burdzy@math.washington.edu

*Real analysis and partial differential equations to DANIEL TATARU, Department of Mathematics, University of California, Berkeley, Berkeley, CA 94720; email: tataru@math.berkeley.edu

All other communications to the editors should be addressed to the Managing Editor, ROBERT GURALNICK, Department of Mathematics, University of Southern California, Los Angeles, CA 90089-1113; email: guralnic@math.usc.edu.

Titles in This Series

856 **Vladimir Bolotnikov and Harry Dym,** On boundary interpolation for matrix valued Schur functions, 2006

855 **Yevgenia Kashina, Yorck Sommerhäuser, and Yongchang Zhu,** On higher Frobenius-Schur indicators, 2006

854 **Noam Greenberg,** The role of true finiteness in the admissible recursively enumerable degrees, 2006

853 **Joachim Krieger,** Stability of spherically symmetric wave maps, 2006

852 **Viorel Barbu, Irena Lasiecka, and Roberto Triggiani,** Tangential boundary stabilization of Navier-Stokes equations, 2006

851 **Jie Wu,** On maps from loop suspensions to loop spaces and the shuffle relations on the Cohen groups, 2006

850 **Siegfried Echterhoff, S. Kaliszewski, John Quigg, and Iain Raeburn,** A categorical approach to imprimitivity theorems for C^*-dynamical systems, 2006

849 **Katsuhiko Kuribayashi, Mamoru Mimura, and Tetsu Nishimoto,** Twisted tensor products related to the cohomology of the classifying spaces of loop groups, 2006

848 **Bob Oliver,** Equivalences of classifying spaces completed at the prime two, 2006

847 **Eric T. Sawyer and Richard L. Wheeden,** Hölder continuity of weak solutions to subelliptic equations with rough coefficients, 2006

846 **Victor Beresnevich, Detta Dickinson, and Sanju Velani,** Measure theoretic laws for lim–sup sets, 2006

845 **Ehud Friedgut, Vojtech Rödl, Andrzej Ruciński, and Prasad V. Tetali,** A Sharp threshold for random graphs with a monochromatic triangle in every edge coloring, 2006

844 **Amadeu Delshams, Rafael de la Llave, and Tere M. Seara,** A geometric mechanism for diffusion in Hamiltonian systems overcoming the large gap problem: Heuristics and rigorous verification on a model, 2006

843 **Denis V. Osin,** Relatively hyperbolic groups: Intrinsic geometry, algebraic properties, and algorithmic problems, 2006

842 **David P. Blecher and Vrej Zarikian,** The calculus of one-sided M-ideals and multipliers in operator spaces, 2006

841 **Enrique Artal Bartolo, Pierrette Cassou-Noguès, Ignacio Luengo, and Alejandro Melle Hernández,** Quasi-ordinary power series and their zeta functions, 2005

840 **Sławomir Kołodziej,** The complex Monge-Ampère equation and pluripotential theory, 2005

839 **Mihai Ciucu,** A random tiling model for two dimensional electrostatics, 2005

838 **V. Jurdjevic,** Integrable Hamiltonian systems on complex Lie groups, 2005

837 **Joseph A. Ball and Victor Vinnikov,** Lax-Phillips scattering and conservative linear systems: A Cuntz-algebra multidimensional setting, 2005

836 **H. G. Dales and A. T.-M. Lau,** The second duals of Beurling algbras, 2005

835 **Kiyoshi Igusa,** Higher complex torsion and the framing principle, 2005

834 **Keníchi Ohshika,** Kleinian groups which are limits of geometrically finite groups, 2005

833 **Greg Hjorth and Alexander S. Kechris,** Rigidity theorems for actions of product groups and countable Borel equivalence relations, 2005

832 **Lee Klingler and Lawrence S. Levy,** Representation type of commutative Noetherian rings III: Global wildness and tameness, 2005

831 **K. R. Goodearl and F. Wehrung,** The complete dimension theory of partially ordered systems with equivalence and orthogonality, 2005

830 **Jason Fulman, Peter M. Neumann, and Cheryl E. Praeger,** A generating function approach to the enumeration of matrices in classical groups over finite fields, 2005

829 **S. G. Bobkov and B. Zegarlinski,** Entropy bounds and isoperimetry, 2005

TITLES IN THIS SERIES

828 **Joel Berman and Paweł M. Idziak,** Generative complexity in algebra, 2005

827 **Trevor A. Welsh,** Fermionic expressions for minimal model Virasoro characters, 2005

826 **Guy Métivier and Kevin Zumbrun,** Large viscous boundary layers for noncharacteristic nonlinear hyperbolic problems, 2005

825 **Yaozhong Hu,** Integral transformations and anticipative calculus for fractional Brownian motions, 2005

824 **Luen-Chau Li and Serge Parmentier,** On dynamical Poisson groupoids I, 2005

823 **Claus Mokler,** An analogue of a reductive algebraic monoid whose unit group is a Kac-Moody group, 2005

822 **Stefano Pigola, Marco Rigoli, and Alberto G. Setti,** Maximum principles on Riemannian manifolds and applications, 2005

821 **Nicole Bopp and Hubert Rubenthaler,** Local zeta functions attached to the minimal spherical series for a class of symmetric spaces, 2005

820 **Vadim A. Kaimanovich and Mikhail Lyubich,** Conformal and harmonic measures on laminations associated with rational maps, 2005

819 **F. Andreatta and E. Z. Goren,** Hilbert modular forms: Mod p and p-adic aspects, 2005

818 **Tom De Medts,** An algebraic structure for Moufang quadrangles, 2005

817 **Javier Fernández de Bobadilla,** Moduli spaces of polynomials in two variables, 2005

816 **Francis Clarke,** Necessary conditions in dynamic optimization, 2005

815 **Martin Bendersky and Donald M. Davis,** V_1-periodic homotopy groups of $SO(n)$, 2004

814 **Johannes Huebschmann,** Kähler spaces, nilpotent orbits, and singular reduction, 2004

813 **Jeff Groah and Blake Temple,** Shock-wave solutions of the Einstein equations with perfect fluid sources: Existence and consistency by a locally inertial Glimm scheme, 2004

812 **Richard D. Canary and Darryl McCullough,** Homotopy equivalences of 3-manifolds and deformation theory of Kleinian groups, 2004

811 **Ottmar Loos and Erhard Neher,** Locally finite root systems, 2004

810 **W. N. Everitt and L. Markus,** Infinite dimensional complex symplectic spaces, 2004

809 **J. T. Cox, D. A. Dawson, and A. Greven,** Mutually catalytic super branching random walks: Large finite systems and renormalization analysis, 2004

808 **Hagen Meltzer,** Exceptional vector bundles, tilting sheaves and tilting complexes for weighted projective lines, 2004

807 **Carlos A. Cabrelli, Christopher Heil, and Ursula M. Molter,** Self-similarity and multiwavelets in higher dimensions, 2004

806 **Spiros A. Argyros and Andreas Tolias,** Methods in the theory of hereditarily indecomposable Banach spaces, 2004

805 **Philip L. Bowers and Kenneth Stephenson,** Uniformizing dessins and Belyĭ maps via circle packing, 2004

804 **A. Yu Ol'shanskii and M. V. Sapir,** The conjugacy problem and Higman embeddings, 2004

803 **Michael Field and Matthew Nicol,** Ergodic theory of equivariant diffeomorphisms: Markov partitions and stable ergodicity, 2004

802 **Martin W. Liebeck and Gary M. Seitz,** The maximal subgroups of positive dimension in exceptional algebraic groups, 2004

801 **Fabio Ancona and Andrea Marson,** Well-posedness for general 2×2 systems of conservation law, 2004

For a complete list of titles in this series, visit the
AMS Bookstore at **www.ams.org/bookstore/**.